MAC

SAGE (Semi Automatic Ground Environment) consoles, 1958.

Former Secretary of Defense
Donald Rumsfeld on his aircraft
in Saudi Arabia, 2003.

SANFORD KWINTER

FAR FROM EQUILIBRIUM

ESSAYS ON TECHNOLOGY AND DESIGN CULTURE

EDITED BY
CYNTHIA DAVIDSON

Author
Sanford Kwinter

Editors
Cynthia Davidson
Michael Kubo (Actar)

Graphic Design
Reinhard Steger (Actar)

Image Editorial
Daniela Fabricius

Image Rights
Cristine LeVasser

Copy Editing
Mimi Zeiger (Actar)

Digital Production
Oriol Rigat, Carmen Galán

Printing
Ingoprint

Printed and bound
in the European Union

Publisher
Actar
Barcelona / New York
www.actar.com
info@actar.com

ISBN
978-84-96540-64-4

DL
B-50943-07

Distribution
Actar D
Roca i Batlle 2–4
08023 Barcelona
T +34 93 417 49 93
F +34 93 418 67 07
office@actar-d.com
www.actar-d.com

Distribution USA
Actar Distribution Inc
158 Lafayette St. 5th Floor
New York, NY 10013
T +1 212 966 2207
F +1 212 966 2214
officeusa@actar-d.com
www.actar-d.com

To my Father, Morris
who taught moderation and practiced extravagance

For Daksha and Sachin

The Mechanical Bride (Stripped Bare by her Bachelors)

Roman Noir

The New Immanence

Space without Qualities (Post-Fordism)

Cobble Stones Beneath the Beach ...

Vortex

Architecture and Air

Rational Irrationalism

Animal Machine

Half of the essays in this book were written for a sounding-off column in *ANY* Magazine provided to me by Cynthia Davidson through the 1990s. The column was entitled "FFE," and while Cynthia and I knew perfectly well that it stood for *far-from-equilibrium*, we occasionally filled out the acronym on the column's title page with the proverbial project architect's more bland and bureaucratic cover phrase: "Furnishings, Fixtures and Equipment." Curiously, during the years I was writing for Cynthia and *ANY*, it was never once confirmed to me that the three uppercase letters of the magazine's title stood for "Architecture New York." In those years in which the typhoon of globalization was sweeping through every cultural rank, the lettering both did and didn't stand for the city in which the journal and the books were published.

These were, it turned out, strange years of change in several uncertain directions, and ambiguity was one of the only clear spaces of intellectual freedom one had to exploit. The *ANY* project, for example, was avowedly internationalist, and yet it still clung inevitably at times to its native tradition, where architectural opinion had for long been generated along the rapidly vanishing axis between New York and Venice (Italy). Despite these preexisting historical and even ideological allegiances, Cynthia never hesitated to encourage and cultivate the irreverence and mischievousness of FFE and never once censored or intervened to bowdlerize in any way. Writing for Cynthia in this way introduced me to a type of pleasure that I never expected to associate with the duty of writing. Indeed, through the 1990s I believe that Cynthia may well have provoked and consolidated the writing habit in me that most reflexively comes to mind when I call myself a writer today. This book was edited by her; it is also a product of the intellectual scene that she sustained and led throughout the 1990s, when so many of the institutions of design thought and speculation had run aground out of exhaustion, conceptual bankruptcy, and cowardice. The final shape of this book is owed not only to this past collaboration but to the delicate and relentlessly cheerful editing touch that Cynthia committed to the book over the past year. This book would not exist but for her efforts.

ACKNOWLEDGMENTS

I owe another significant but invisible debt to the clusters of thinkers informally grouped throughout the still-forming international scene of the 1990s, to the inchoate alliances at first grounded in friendships alone but which later matured into mutually inflected intellectual confederacies. In this category, the sustained conversation with Jeffrey Kipnis that was initiated in Los Angeles, and which then gathered steam in London, was among the most dynamic and constitutive. His courage, in recent years, to modify his practice from teacher to curator, to writer, and then to a thinker of risk and vulnerability (allowing his own passions as musician and physicist into his architectural thought at first purely as a favor to me), have provided sustained inspiration and awe. I would like to acknowledge as well Robert Somol, who at times may have been my only reader, yet whose wickedly hilarious responses made having a single reader pleasure and justification enough. Albert Pope was a steady and rigorous interlocutor throughout the FFE years, the kind one needs when going out on a limb routinely without checking branches or landing spots. Lars Lerup provided an unusual atmosphere of official encouragement for risk-taking and a natural love of ideas that is rarely so actively promoted for its own sake in architectural schools. His example as a writer has provided a rare and important opportunity for speculative collaborative thinking. It was his continual insistence on the City as the litmus test of legitimate architectural speculation that lies incorporated in the foundations of the present book. Indeed, the entire faculty of colleagues at Rice—John Casbarian, Will Cannady, Carlos Jimenez, Albert Pope, Gordon Wittenberg, Fares el-Dahdah, Christopher Hight, Spencer Parsons, Dawn Finley, David Guthrie, Nana Last, Clover Lee, Sean Lally, Doug Oliver, Stephen Fox, and Danny Samuels—provided a general camaraderie of encouragement and laughter that made shaking things up both on the home front and abroad a constant source of fun. Among the central "silent" interlocutors of the 1990s, whose creative presence provided the necessary gravity and challenge for real thought, were the architects Rem Koolhaas, Greg Lynn, Jesse Reiser, Nanako Umemoto, and Lindy Roy, and of course, as always, Peter Eisenman. In a cryptic way, it was always their work I was trying to write about.

Intellectual debts in a book of this type are hard to express with due humility. But the debts I owe to the work of George Grant, Marshall McLuhan, and Michel Foucault never found appropriate acknowledgement in the texts themselves, so implicit and saturated are the essays with attempts to reconcile the brilliance of these thinkers' examples. I wish to say that I almost always felt cheered on during the difficult years when my tirades against the mediocrities of digitalization made me a confusing and unembraceable figure for more than one generation of 1990s student. On this front I wish to acknowledge the brave work of Branden Hookway, once a student and now a colleague, with whom I collaborated both on his own book *Pandemonium* (1999) and on the "Dossier" of critical-historical readings assembled at the end of the section of *Mutations* (2000) titled "The American City." His work in many ways has been a necessary and serious support and justification for mine. In this same vein a specific and a general debt of gratitude must be extended to Daniela Fabricius, whose work after her assistantships and collaborations with me have provided a form of encouragement of the type that can only be had from workers of a younger generation who have refused the platitudes—and in many cases the rewards—of capitulation to their own generation's habits of mind. The specific debt to Daniela is for the patient and hard-nosed image editorial work that is an inseparable aspect of this book's attempt to communicate attitudes and ideas both responsibly and playfully. Similarly I would like to acknowledge the loyalty of yet another former student and now colleague, Ana Miljacki, whose unwavering faith over the years, when she was forging her own political attitudes, made the self-imposed contrarianism of the 1990s not only bearable but fun.

I wish to thank certain people for the personal and institutional support they extended, who embraced and participated in controversy as a matter of principle: Bernard Tschumi at Columbia, Marlon Blackwell at the University of Arkansas, Jorge Silvetti, Michael Hays and Mack Scogin at the Harvard Graduate School of Design, Stanley Tigerman at Chicago, Rob Livesay, Jeffrey Kipnis and José Oubrerie at Ohio State, and Alan Balfour and Kipnis again at the Architectural Association in London.

Subsequent support from Mohsen Mostafavi at Cornell, Toshiko Mori at Harvard, Daniel Birnbaum at the Staedelschule in Frankfurt, Detlef Mertins at the University of Pennsylvania, Wolf Prix and Reiner Zettl at the Universität fur Angewandte Kunst in Vienna, and Kent Kleinman, formerly of SUNY Buffalo, provided both a wide platform for my own education and the inestimable gift of solidarity. I would like to especially thank Adele Santos and Yung Ho Chang for the opportunity over the last few years to teach at MIT and interact widely and across disciplines in the inimitable way of that institution, thus giving a kind of post-facto legitimation for thinking of design and technology within the same culture.

I would like to thank Cristine LeVasser for her swift and efficient cleanup of the arduous work of procuring images and permissions. My thanks as well to Jillian Maslow, who patiently lifted much of the burden of competing administrative tasks long enough for me to see this book through its final stages of completion. I would also like to acknowledge the generosity of Franziska Schelle, whose patient and cool efficiency were essential for clearing the time necessary to complete the final details of this book. Others whose general support I felt perpetually at my back are Rem Koolhaas, Sylvia Lavin, Michael Bell, Johan Bettum, formerly of the Henie Onstadt Center in Oslo, Scott Lash at Goldsmith's College in London, Brian Boigon of ThinkThinkThink, Ben van Berkel, Mark Jarzombek at MIT, James Corner, Karl Chu, Jacques Sautereau, the late Sheila Zeldin, whose enthusiasms were a force of nature, and Bruce Mau. I must also acknowledge a great many people younger than I with whom I taught and who taught me so much: David Ruy, Jason Payne, Anna Pla Catala, Reto Geiser, Christopher Hight, Luke Bulman, and Michael Maggio. Roger Conover at The MIT Press contributed valuable detailed input on an earlier version of the book.

I would like to thank Albert Ferré and Ramon Prat of Actar for providing over the years a publishing relationship that twenty years of experience in American publishing (via Zone Books) never led me to believe was possible. Working with them since our millennium book *Mutations* has been one of the most unexpected and exhilarating experiences of my publishing life. The work, commitment and generosity they put into the present book reinvigorated my own interest in the project and supplied an incredible impetus and wit to the final work. I would also like to thank Reinhard Steger, who designed the book with a mixture of restraint, intelligence and audacity. I feel incredibly lucky to have once again been paired with a designer of immense graphic talent who yet still maintains an ultimate respect for the printed word, and even more so, who possesses that rarest thing among book designers, a clear and nuanced *theory* of what a book is.

Michael Kubo drove the manuscript from a pile of papers into the book you see before you, contributed broad control over the flow, concepts, look, and tone of the book, acted as a liaison with the designers, and served as a powerful advocate for the book on every level. Many times he explained to me what parts of my book meant, or could mean, and served as an extraordinary intellectual sounding board for every issue of content and form. I consider him much more a consultant and final publication director than an editor.

Many of the essays in this book first appeared elsewhere, often in slightly more pugilistic form. As a Canadian educated first in the English and then the French tradition, during the two decades after the 1960s I was accustomed to a certain ludic interplay among interlocutors; in other words, a certain ritual of gauntlet-throwing and public challenge and debate. In the American context in which most of these essays were first published, they found almost no direct counters or responses (although there were a number that originated from European contexts). While I have no lasting regrets about this, the placidity was much compensated for by the fact that many of the figures who began as adversaries (albeit straw men) over time ended as good friends and colleagues with whom I came to feel a powerful and permanent solidarity. Perhaps it is from this precondition of affection that more hardy and humorous debates might, in the future, ensue.

Finally, I would like to formally thank those "unknown soldiers" who wrote spontaneous personal letters of support for positions I took during the dot-com era that teetered— at least by the standards of the time—on "fringe" expression. There were letters from crackpots to be sure, but also from surprising places where one certainly would not have expected to find the rudiments of an "underground" resistance, such as from workers at Xerox PARC, Apple Computer, certain prominent art museums, and even from administrators in public psychiatric medicine. These treats are perhaps among the secret pleasures of doing criticism, and I don't mind admitting that their importance to me grew over time.

No litany of thank-yous could omit the generous proposals of assistance extended to me spontaneously over the years by the Graham Foundation for Advanced Studies in the Fine Arts, initially by Richard Solomon and finally by the indomitable Sarah Herda, the Graham's current director, who finally converted the earlier offers of principle into a concrete reality. Had she not stoked the motor of this manuscript with her own incredible capacity and impatience to make things happen, this book would almost certainly still be languishing in a "to do" pile in the most desperate corners of my office. This book's final push to completion, and my ability to escape from a great many aspects of its drudgery, is a product of Sarah's and the Foundation's support.

In closing, I would like to acknowledge the near infinite debt I owe to Gita Vaid, who, despite her belief that I make everything up in my sleep, nonetheless grants me occasional moments to think things up while awake. Gita's life and work (as a psychoanalyst) consists in discovering patterns in the speech and actions of others, which means that her capacity for forgiveness and for tolerance must grow with every revealing idea or text I produce. If not for her incredible capacity to laugh it all away at will, to turn a blind eye to the pathologies from which I can no longer even hope to escape, I suspect that the vocation of writing in the context of family life would not be possible at all. My debt to her for the forbearance required to tolerate travel and the slow work of writing, as well as for bringing our two gorgeous boys into the world, is inexpressible, but here at last it is registered in print.

. . . encore de l'audace, toujours de l'audace!
Georges Danton[1]

The phrase *far from equilibrium* is borrowed from the field of thermodynamics where, especially in the last four decades, it has come to refer to the special states of a system in which it is most likely to produce radical, productive and unforeseeable behaviors. When close to equilibrium, the disturbances, anomalies and events passing through a system are easily absorbed and damped out; but as a system is moved further from its rest places it becomes increasingly ordered and differentiated. The more ordered and differentiated, however, the more unstable and expensive (in terms of energy required to sustain it) the system becomes. Instability, it turns out, is the precondition of creativity. But because of its cost—in terms both of the energy required to produce and sustain it as well as in terms of the damage to the status quo that must invariably be repaired following every transformation—instability and invention are rare.

Among the abiding mysteries of contemporary knowledge—in addition to the workings of the brain, of evolution and of Nature itself—is the mechanics of history and historical change. It is simply not known how history works, why change is constant yet not random. Causality has often proved to be a good myth to live by, but no more than that. We accurately think of ourselves today not only as citizens of an information society, but literally as clusters of matter within an unbroken informational continuum: "We are all," as the great composer Karlheinz Stockhausen once said, "transistors, in the literal sense." We send, receive and organize, and when no longer capable of sustaining these three functions, we disintegrate and die. But so long as we are vital, our principle work is to capture and artfully incorporate the signals that surround us. Life is a case of maintaining a very delicate structure—ourselves—a significant distance from equilibrium at nearly all times, and at others—in order to evolve, grow and invent—very far from equilibrium indeed.

In common parlance, equilibrium and instability have opposite associations: equilibrium is calm and steady and sure, instability is danger and impurity itself. The title of this book has its purposes served here as well. The pieces were written in a deliberately fervent and essayistic style: fully affirming the subjective nature of the writer's position, and intending to provoke, at all and any cost. The field of design has two principle venues: the first is the public

Introduction
DE L'AUDACE . . .

sphere of mostly promotional press, and the second, the much more insular sphere of academia. Each in their own way has capitulated utterly to the weather of the crowd and to a false and shallow collegiality. At any rate, they have become allergic to controversy, at times embarrassed even by humor, mild mischief and anything more than the most ironic irreverence. Irony is not something the reader will find in these pages. A third venue of design, although in a dangerously atrophied state at present, is the field of criticism: the public, it seems, has no further tolerance for having its objects of pleasure and magic submitted to anything more than aesthetic or physical scrutiny. For every object conveys a world of hope and carries within it the best hope that can be found in the world that made it. Should this hope be found to be wanting, or mediocre (or worse), it is not the object that will bear the brunt of the deception, but those whose world and phantasms depend on the persistence of the veil of a now broken magic. At bottom, I believe that design is a highly advanced form of rationality, perhaps the highest there currently is, at any rate one that has surpassed art in so many ways in our era, and that it is a legitimate vehicle of the deepest human strivings, foremost among these, of hope. In this sense, more than just the congenial flow of our living rooms and silverware depend on a clear and unsentimental—independent and audacious—commentary and discourse about the designed environment. Far from equilibrium denotes for me a duty to write from the very fringe of coherence (the false coherence fostered by consensus), the duty to actively remember, and the duty to imagine what is not being said and say it as if it were real and not only invented. To do this, one has to make an ethos out of the far from equilibrium state. It is to this that the title and epigraph above refer.

If design is the dominant form of rationality in our era, it is inseparable from the grand machinery of secular striving and making identified by Max Weber a century ago; it compounds our economic, spirito-religious and sensual life into a single yarn: it is *technique* itself. To say that it is what we are, is not necessarily to celebrate, but to cast a warning and an admonition that somewhere the control of our destiny was handed to us and we failed to answer the challenge with either sobriety, ecstasy or thought.

The writer Marshall McLuhan liked to point out that of all the modes through which the human organism is endowed to participate in the world, only that of vision and its corresponding apparatus, the eye, has the capacity to focus.

THE MECHANICAL BRIDE
(Stripped Bare by her Bachelors)

Our other senses are panoramic, prodigal, continuous and integrated; they overlap, seep into one another's synapses and always deliver a world that is single and woven—an organic totality. The capacity to isolate, to intensify one element in a field at the expense of the others, to establish hierarchies by actively molding and truncating attention, lies at the root of the dissociative habits that make our modern culture of *analysis*—and its linear, sequential processes—possible. Yet, because the senses always work together (historical unfolding, and its elaborate network of technological cues, can change only the mixture or ratio of the ensemble through which the senses operate), and because no voids are possible in an attentional field (the nervous system, too, abhors a vacuum), the massive set of reshaping processes that characterize our world, and the often brutal deficits that accompany them, always and necessarily remain hidden from us.

We cultivate and theorize our technological environment today in strange and partial ways, without ever admitting to ourselves that this is what it is, *an environment*. We defend many of our new, especially media, technologies with a blind, implicit, and often mindless zeal for a type of "progress" which would have been routinely dismissed only a couple of decades ago, while one-dimensional theorists continue to spearhead an ironic—though thoroughly unconceptualized—historical reversal: to emancipate human spirit, imagination and energy from the tyranny of economic rationality no longer through the values broadly associated with moving "back to the land," but through an even more adamant anti-urban migration into the military-industrial-bureaucratic (to which one might well today add "-entertainment") landscape itself, into the synthetic, Balkanized, narrow-bandwidth ghettos of the "nets." Nothing is more laughable, or dangerous, today than the continuing fatuous belief in the new freedoms offered by the "closed city" of Internet sociality.

We continue to hear the fiction that what is important is how a technology is used; we are sermonized on the—microscopic, short-term, immediate—advantages that such technologies claim to offer us, but hardly ever do we meditate on wider implications: on what a given technology, or group of technologies, by virtue of nothing more than their simple presence within our perceptual economy or Lifeworld, actually does *to* us.

The disciplinary apparatus that philosopher Michel Foucault once described[2] no longer remains the familiar trope it thereafter became, because the mechanisms through which its active logic is being embodied in our current society—refashioning us as human beings—are mutating once again. In Foucault's time, "humanism" was justifiably excoriated as a prime enemy of free action, mainly because it fostered a narrowly confined, yet "total environmental" notion of what the possessor of freedom—the "subject"—ought to be shaped like. This environmental effect lies behind Foucault's concept of "productive" power, the power that coerces without observable violence and without the disruptive dissonance of an imperious "no," and it also accounts for why most of the absences and deprivals in our social and historical world—no matter how fundamental—have a way of becoming instantly indiscernable to us. In the face of sweeping new historical developments such as the assembly and imprisonment of human subjectivity within, and its transfer to, purely engineered and machinic (software-driven) environments, much that was once rhetorically—and negatively—connoted by the term humanism may well today be subject to a potentially new and radical revival. Resistance to the insidious de-urbanizing pressures of the new electronic disciplinary apparatus—such as through acts of radical anamnesis and new ways of cultivating archaic modalities and intensities—must today courageously follow lines of thought and action unimaginable to the poststructuralist and postmodernist mind of the late twentieth century. One way is to conceive of human subjectivity, the body it is formed in, the ecstasy it is capable of, and the trajectory it is moving along, within a broad ecological model still sensitive to the untapped revolutionary possibilities that remain enfolded within past worlds and objects. If nothing else, such a model would at least permit us better to see and to judge the unprecedented mediocrity of our present aspirations, and indeed as well, the possibly dire importance of what we currently seem all too willing to give up.

It is hoped that the present book will serve as an example of what Michel Foucault once described as a decisively missing aspect of much 1960s militantism: the insight that agitation was never meant not to be fun.

1 "What is needed is audacity, and again audacity, ever and always audacity!" This famous phrase of French revolutionary Georges Jacques Danton was uttered in September 1792 on the eve of an invasion by the Austrian Army and two days before being elected as Deputy of Paris. He was beheaded less than two years later. His last comment to his beheader: "Don't fail to show my severed head to the people . . . it is well worth seeing."

2 Michel Foucault, *Discipline and Punish: The Birth of the Prison*, trans. Alan Sheridan (New York: Pantheon Books, 1977).

It is arguable that no fantasy marked twentieth century aesthetics—or cultural Modernism itself—more powerfully than that of the Bachelor Machine. Although the term was coined by Marcel Duchamp to describe the web of forces brought together to animate his seminal "Large Glass," the phenomenon of automated desire—the desire, that is, whose origin is to be found in the social field rather than in the individual person—began to emerge in the late ninteenth century and has mushroomed ever since. The idea that the technical realm *is* the realm of (social, political) desire was taken up from the beginning by Diller and Scofidio in a sustained set of enactments and developments of the ideas of Duchamp that were manifested in performances, installations, videos, objects, and texts. The logic that determines our routines of inhabiting the world were seen to exist at the level of buildings only in the most secondary of ways, and so for long their practice explicitly eschewed "Architecture" in every classical sense of the word. In time, their researches transformed architectural thought by providing a more sophisticated understanding of the how the design of the environment in general determines who and what we are. Primary among their interests, as for Duchamp and McLuhan before them, is the role of the eye as a shaper of space and of social and psychic organization. With the historical rise of the cinema and its constitutive role in training modern perception and assigning a new interrelationship of the other senses, the meaning and uses of "space" were transformed in turn.

Diller Scofidio + Renfro, perhaps more than any other practice, have shown the primordial importance of the concept of the "program" as the determining factor in the lived environment. Adopting what one could call the "paranoid-optical" model of modern civilization, Diller Scofidio + Renfo relentlessly invent—or extract—vision machines within and out of every cultural nexus or material locale: every room, building, city, or site is revealed as a machine that distributes visibilities and invisibilities like a perpetual electromechanical striptease in which humans are the invariable prey caught in an ecstasy of self-watching (television unsurprisingly serves as their principle rhetorical "device").

The Blur Building depicted here is naturally no building at all, but a desiring device of architectural scale and scope. The mist produced by the 31,000 nozzles that artificially atomize the natural lake beneath it generates a veritable *Ganzfeld* in which the body, for lack of a perceivable fixed point in the sensible environment, is unable to orient itself in space (or time). The vertigo and disorientation that ensues is intended to go beyond mere physiological effects, leaving the spectator-inhabitants trapped in a state, not of blindness, but of total self-perception, seeing sight itself unencumbered by any object. The total immanence of the experience becomes on the one hand a type of routinized Enlightenment (the oneness of the perceiving self and the world perceived), and a wicked revealing of the invisible "apparatus" from which we can already no longer escape.

The number of cell phones anually discarded in the United States alone is 130 million.
To display all 130 million would require 43,000 more images identical to this one.

The Kazam! machine in the Eameses' Los Angeles apartment, 1941. The first apparatus for the controlled bending of plywood, adapted from the shipbuilding industry and later serving as the model for the proverbial and omnipresent "NURB."

PLAYBOYS OF THE WESTERN WORLD

You're so vain, you probably think this piece is about you...

It has certainly never been more difficult than at the present to maintain a sincere fervor or romantic attitude regarding the institution of the museum, the special objects of its custody, or the social, psychological, and sensual encounters that it organizes around them. Since the end of the ninteenth century—that is, since the rhetorical formations of the avant-gardes—art ceased its explicit function as the shaper of the conventional tastes and values of a conservative and primarily "landed" class, to become that class's direct antagonist and predator, the undoer of the very values by which mere ownership could be converted—with interest accrued—into right, power, and entitlement. Yet as Friedrich Nietzsche pointed out, aesthetics has always been a form of warfare—at first, valuations represent the trophies of successful, dominant and sanguine "Wills," later, the mere spoils of crude and defensive vested interests. The arrival of the museum concept, indeed of the tradition of maintaining cultural canons in general, may well be the threshold that marks the passage between these two regimes; custodianship was simply never the activity of a creative and robust spirit but that of one in retreat, one with something to lose. The original museums would protect bourgeois Enlightenment ideals (whether interesting or not) from the natural predation of historical becoming (whether progressive or otherwise). Today, however, it is historical becoming itself that the museums claim to protect, and they pose no longer as custodians of dead artifacts, but as mavens and cultural producers, self-appointed purveyors of the social-aesthetic imagination. In light of this new, and thoroughly proclaimed, constellation of functions, it does not seem unwarranted today to ask, "What set of political or desiring instruments should we adopt to protect ourselves from the museums?"

The vagaries of democratization are especially tricky today. Since the Centre Pompidou's inauguration in Paris nearly thirty years ago, the museum has achieved a level of spectacle of truly unprecedented magnitude, immense in its capacity to draw crowds, to generate (and expend) income and to manufacture prestige for visual culture. The image industry, of which television, advertising, the cinema, and the shop window are only the most self-evident examples, belongs unquestionably to a set of processes whose scope has become almost dimensionless, yet whose invasive intimacy borders on a kind of micro-violence.

First published in *ANY* 13, "Tate Frames Architecture" (1996), a special issue on the Tate Modern design competition.

The Hibernia oil platform off the coast of Newfoundland is half the height of the Empire State Building and accommodates a crew of 185.

Does art continue to protect us from these now routine predations, as once, through its abstractness, difficulty, rarefaction, and remoteness, it formerly did? To certain invertebrate theorists the demise of the "classist" distinction between high and low culture is an evolutionary process worth abetting and celebrating; "popular" has come to mean nothing more abiding than "pop," and their own perorations on the subject have about as much originality and demand on cultivated attention as the word's abbreviated sound has on the ear. The art game, married more deeply during the 1980s to commerce and capital than at any time in modern history, is now an indistinguishable, and possibly indispensable, relay within its inclusive system. The democratization of art, as if art's rarefied intensities were just another form of political privilege among others to be done away with, was thus achieved through the industrialization of aesthetic production and dissemination. The industrialization of art—of which the museum is the primary rationalizing locale—was achieved, however, only in the wake of an earlier process of reduction, the reduction of artistic predation and practice to the mere sign of its logic or identifiable contents: the reduction of art to image.

Not only does art—and the formal, ritualized structure of our approach to and encounter with it—no longer protect us from anything, but, through the multiple agency of the museum and the media networks with which it is fused, art has now fashioned and targeted us—metropolitan citizens—as its prey. Of course, there is no end to the celebrations of the "new modalities of artistic production and reception" that are on the horizon, and the words sound right, but the ways these are understood by the citizen-artist-spectator and the institutional custodian-impresario-adman, ought really to be radically distinct. When a major new museum, then, is not only built but, in fact, reconceived, and with a consciously philosophical view to addressing the future ways that humans will address themselves to the things they make, it must be realized that a critical political opportunity has presented itself. After all, art itself is not "important." (It is primarily museology that invests importance, and then only as units of capital under the manifold pretext of a "history" that requires preservation.) What is important are the relations a society permits its actors to maintain with art. What could these relations be, and how can architects respond to this massive gap in culture, the psycho-social and political space abdicated by the artistic avant-gardes?

Alas, much of the architectural profession is in no position to preach to anybody. Yet as if by some astonishing clairvoyance, by giving priority to research and relationship over scheme, the organizers of the 1994 Tate Gallery Design Competition in London appear to have overcome the all too common thrill-seeking reflex to the empty spectacle and the pyrotechnics of form—to have rejected the apparatus of the desperate playboy seduction in favor of the sustained peripatetic relationship of reflection, sustained transaction, and, as the Greeks might have put it, the journeyworld of sublimated love.[1]

Design competitions are marvelous and necessary institutions: they encourage innovation and research, speculation and brave experimentation. They are the closest thing to a true laboratory freedom that the profession is able to offer itself, and it is the means by which a great deal of architectural history is made, if only because the easing of "real" constraints permits and encourages a fluid and uncompromised pursuit of ideas. And yet, the mediatization of architecture has invaded the competition laboratory just as fully as it has the artist's studio. The peripeties of the "idea," the subtle movements and narrations of intricately converging and speculative processes are lost in the contraction to the circulatable, instantaneously digested image. What has overtaken architecture today, though by no means destroyed it, is the same practice Sigfried Giedion described fifty years ago: the mindless "hunting for form" that characterizes "playboy architecture."

As a morphologist of sorts, I must accept what responsibility is my due. I am certain, however, that I have never discussed form as an end in itself, but only as the distant expression of a complicated set of interacting dynamics whose roots are nowhere else but in the social, historical, material world in which our own human drama unfolds.[2] I am often surprised at how architects, both those brilliantly deft at manipulating form and those dismally ungifted with it, equally defend the position that form is the only rightful end of architectural propositions. I am shocked at how egregiously silly they sound and at how unaware they are of it. The "hunters for form" take many guises and cut across all generations: young narcissists stricken with computer puppy love, who believe in magic (and software faeries too no doubt), and are totally enthralled with whatever the beloved machine digitally excretes; the rodeo gang with big offices, big personnel, and military software ("no time for thinking here boys, just aim . . . and shoot"),

Intercept monitor room at the SAGE Direction Center at MIT's Barta Building, Cambridge, MA, 1955. This room served as the prototype not only for the future national air defense system, but by the 1990s, for every room in the developed world. These behemoth machines had the computing capacity of a cellphone.

who favor anything gymnastic that other, smaller operations can't manage (and which fact alone, they think, merits them the commission); the eight-dollar-word boys who clearly know a thing or two, who spin a fancy, provocative phrase but deep down know that bimbo structures—cosmetic silicone and aerobicized drawings—are what's really turning heads these days; finally, even octogenarian wannabees, who rightly and wisely deny software's significance or power for formal understanding, yet offer the world trivial Form-Z stock artifacts tackily gift wrapped with campily mispronounced titles while fashion-victim critics slavishly bleat "masterpiece" behind them.[3] The hunters for form are everywhere, and almost nowhere are they discouraged.[4]

The sadness of mere form is a visceral one, inescapable, like the *tristesse* of an aborted erotic encounter. The current hyperbolization of mere image-form, a kind of weird necro-glamour at the *fin de siècle*, has given place to an unbelievably powerful renewed intuition that is already sweeping the playboys into oblivion and undoubtedly will continue to do so in the years to come. That intuition, whose first version marked the dawn of this century and the very agar of the avant gardes, consists in a truism so unsettling it has taken a century to grasp: that form and architecture as such is no longer interesting, indeed can no longer make the slightest historical claim on our attention (and no CAD package can change this fact). What is interesting today, and what matters, is *infrastructure*.

Although modernists offered this insight long ago, only today can we see that it was actually prophecy. The playing field has changed, but has done so decisively only recently. Infrastructure must be understood as a broad term, a new, largely invisible type of materiality, much too big and serious and delirious for the magazine gladhanders and the playboys with their desktop gadgets and animated pictures and special effects. Infrastructure is big, ubiquitous and foundational: It is the systemic expression of capital, deregulated currency, interest rates, credit instruments, trade treaties, and market forces; it is water, fuel and electrical reservoirs, routes and rates of supply; it is demographic mutations and migrations, satellites and lotteries, logistics and supply coefficients, traffic computers, airports and distribution hubs, cadastral techniques, juridical routines, telephone systems, business district self-regulation mechanisms, evacuation and disaster mobilization protocols, prisons, subways and freeways and their articulated connections, libraries and weather-monitoring apparatuses, trash removal and recycling networks, sports stadiums and the managerial and delivery facilities for the data they generate, parking garages, gas pipelines and meters, hotels, public toilets, postal and park utilities and management, school systems and ATMs, rail nodes and networks, television programming, interstate systems, ports of entry and the public goods and agencies associated with them (INS, NSA, IRS, FDA, ATF), sewers and alarms, multi-tiered military-entertainment apparatus, decision engineering pools, wetlands and water basins, civil structure maintenance schedules, epidemiological algorithms and virology labs, cable delivery systems, law enforcement matrixes, licensing bylaws, green markets, medical-pharmaceutical complexes, Internet scaffolds, handgun regulations, granaries and water towers, military deployment procedures, street and highway illumination schemas; in a word, *grids* of any and all kinds.

Exfoliating the grid:[5] this is the design task that history offers us today. Architecture as we once knew it, the parlor game of the private commission, the economics of the elegant building, will form at best, tomorrow's doggerel. Today's playboys of course will become tomorrow's Deadheads, crowding around the quaint computer screens of the twenty-first century, cooing "wow" as the 395th glorious variant of the same digital meatball renders out in four splendid dimensions. All around them a world is being designed, in a sometimes sinister, sometimes beautiful manner, but none of this concerns them in their assiduous and tireless hunt for form.

The exfoliators of the grid on the other hand are designers who work from research, hypothesis and experience, who are guided by ideas, who cultivate the ethos of the soft hacker, who are connected politically to their actions, and who seek to account socially for their consequences. The exfoliators of the grid are thinkers who know how to amplify their gestures to macroscopic, sometimes even global scales; they think and act boldly even when working in miniature, because they work within the material of the grid itself, not the sealed suburban alembic of the playboy's formal reductionism. They know that the age of creation *ex nihilo* is over, that it remains only as a sepulchral residue, the foundation of playboy metaphysics.

The Tate Gallery Design Competition brief showed all the signs of wishing to call this hermetic function out into the open. Optimistically, one imagined it no accident that a power station was chosen to anchor the museum/urbanism dyad project. Infrastructure simply has an aura different from that of building: it possesses a critical resonance, a direct reach or extensivity into the adjacent material environment that no architecture alone can approximate. Infrastructural assemblages operate like nodes in sensitive, ductile tissue, patches of an active—or at least activatable—*grid*. In other words, infrastructure is already *intelligent* architecture; it represents a concrete *metaprogram* already communicationally embedded within other architectural manifolds or continua, poised and ready to be switched on. Many of the Tate finalists seemed clearly to espouse this insight: David Chipperfield's brilliant conceptual scheme of a miniature econosphere embedded, like a microprocessor in a motherboard, within London; Michael Hopkins's underappreciated indoor urban "station" or phantasmagoria à la Ferris or Sant'Elia; Rem Koolhaas's *cours des miracles* artpark assembled entirely according to planning rather than architectural principles; Renzo Piano's quiet but bold clarification and extension of the building's bass-note spaces and affects to emancipate soft plateaus and cavities like civic gallerias, piazzas or squares. With such a bold field one is left only to wonder how a project as trivial as the one submitted by Rafael Moneo's office survived to the final round. The supporting text[6] offered its own fatal indictment: nowhere was there an idea, intention, or perhaps even capacity to formulate a problem broader or more worldly than that of solving an elegant entryway to a building conceived as no more than a group of tectonic *faits accomplis* in need of remediation. This was not even playboy architecture, but royalist architecture of a most troubling kind. Were the Tate judges cowed by reputation, or is it simply wrong to credit them with possessing the same clarities of purpose and concept as evidenced in the "call for entries" document drafted by Tate Director Nicholas Serota? Or perhaps one is simply reading too much earnestness into the whole museum affair and its historical and infrastructural implications to begin with.

In the final analysis, it is no surprise to any observer of the European scene in the 1990s that Herzog & de Meuron were selected to be the architects of the Tate's Bankside extension. Europe (the new Euro Europe, that is) has long been intoxicated with the Modern Look, the taut, crisp, lean, luminescent structures of Jean Nouvel, OMA, and of Herzog & de Meuron.

Tate Modern, Herzog & de Meuron.

But Herzog & de Meuron seemed to have carried a special appeal. On one level it was the appeal of light itself, though it was little more than that of the overexposing cathode ray tube etching its characteristic white hole into a dark room (an aesthetic celebrated in the happily lite "Light Construction" show at the Museum of Modern Art in 1995, and which served as the final lullaby for architectural ideas and exhibition in New York for what is approaching nearly a decade).[7]

As the video image in the Herzog & de Meuron scheme burns, the worldly details fade; the resultant vagueness is transmuted into an effect meant to resemble mystery, and events transpiring within the building are transmitted and recorded at the surface radically, incompletely, though in an always highly stylized way: the architectural correlate of the blasé valley girl's "what-*e*-ver." An atmosphere of sumptuous poverty is scrupulously developed. Evacuated spaces are counterpointed with highly activated yet never hysterical surfaces where communicational intensities are always only partial, adhering strictly to what linguist Roman Jakobson once called the "phatic function," the flow of ums and uh-huhs that keeps interlocutors online during conversational lulls when no significant information is being exchanged. These visible phantoms and shadows tantalize one with the vaguely erotic promise of a solid encounter, yet such encounters are never scripted in—in such an exquisitely choreographed world nobody seems to want one any longer. It is the pure, serene, affectless mirage of electronic intercourse upon which so much of contemporary subjectivity—to wit, MTV, the Internet, etc.—is founded. In true daylight, Herzog & de Meuron's dematerialization strategies assume a different, but equally sophisticated tack. Surfaces claim informational density, not in the direct sensual terms that would be dictated by their actual chemical-physical compositions but in their very capacity to retreat behind the images they carry. Ceilings, walls, floors, columns, patios and doors are overcoded like so many palimpsests with poignant, if banal, images that are repeated serially to produce a carpet of suggestive but vacant totems on an epic scale. The effect is certainly sublime . . . for an instant, at least.

Yes, Herzog & de Meuron may well have become the rightful architect-poets of our era, solidly in touch with the weary, disengaged logic of the contemporary world—grunge mind plus Swiss bank account in nightclubber's garb—unconcerned, and for the most part unaffected, by the boundaries of where the logic of image and "spectacle" ends and where the "situations" of the fulminating life of the human civilized savage begins. The ethereality and remoteness of their work, its quiescence, should be mistaken neither for neutrality nor for a Zen-like focus or stillness. The work is both severely polemical and hyperactive in its narcissism: it is both mirror and image, the one frantically searching for the other in a *drague macabre* across every surface, a surface which is merely a neat, clarified, hygienic extension of that of the fraying, hypermediated outside world. For anyone still gullible enough to believe in the efficacy of negative dialectics, the Tate and all the playboys of the Western world have undoubtedly got themselves just the building they deserve.

1 It is, of course, not abstinence that is being advocated here, but rather a refusal of the ethos of the premature ejaculation, or the tawdry five-minute fix.
2 See "Who's Afraid of Formalism?" (this volume, page 144).
3 Philip Johnsons's sophomoric New Canaan gatehouse, which he referred to as the "monstah," was first exorbitantly presented by Herbert Muschamp to *New York Times* readers as a "masterpiece."
4 America's preeminent magazine of highbrow design culture in the 1990s—*ANY*—became itself an altar on which intelligent, innovative design was sacrificially slaughtered in the service of cataloguing every ill-digested graphic design technique and recycling every vacuous typographic experiment of the previous six years. The result was a sometimes genuinely seductive total image at the point-of-purchase interface, achieved—at a cost—only through a posture of philistine contempt for, and an assault on, the intellectual content—and life—of its articles. If the medium became the message in the modern, post-phonetic world, then in the 1990s, it often seemed, the message had become the pure media freefall itself. Of all forms of reductionism, the reductionism to form and image is perhaps the most pernicious of all.
5 The term "exfoliation" is borrowed from the topological work of Fields Medalist mathematician René Thom. Exfoliation refers to the act of forcing an expression (or geometrical unfolding) of forces present in a field which have not yet been triggered into visibility.
6 Published in *ANY* 13.
7 "Light Construction," curated by Terence Riley at The Museum of Modern Art, September 21, 1995 to January 2, 1996. See Terence Riley, *Light Construction* (New York: The Museum of Modern Art, 1995).

Sonic Boom. The formation of shock waves behind the jet produces the sharp drop in temperature that leads to the condensation of water vapor and the half-anticipated halo of angels behind.

MACH 1
(and Other Mystic Visitations)

Architects do not lack for work these days, yet still they fret, and it is small wonder. Powerful and disturbing millennial forces are everywhere afoot and their failure to be adequately discharged or convincingly absorbed is making mincemeat of the perennially nervous profession. The prestige of architectural ideas and practices, about to achieve self-confidence as it rises to a par with that of other arts and disciplines for the first time since the Renaissance, is again having its thunder stolen almost daily. The efficient but one-dimensional marketplace world in which we live, fraught to the exclusion of almost everything else with relations of exchange and driven by forces of unrelenting supplication, is a world, one could say, in which we are hectored mercilessly by design, swathed in its miasma of artificial affectation, hyperstyle and micro-human-engineering that anticipates, like a subtle reflex arc, our every move and gesture. Design has now penetrated to, even threatens to replace, the existential density, the darkness, the slow intractable mystery of what was once the human social world. In its place it is installing a shrill new immediacy, a garish transparency of postures, glances, and address. Design has become us; it is, alas, what we are, and there is no way (for now at least) to separate ourselves from it. Even formerly lumpen cultural sectors such as the commercial fashion world now exercise a grip on minds and sensibilities, postures and moods, producing figures of genius (Rei Kawakubo, Martin Margiela, Hussain Chalayan), and heroes of popular notoriety (Alexander McQueen, John Galliano and the consumer house-cults of Prada, Chanel, and Gucci) from within a system of dissemination so vast and inclusive that architects would not even know how to dream of competing. Amid the shameless ersatz novelty of the fashion world there are flickers of true innovation—and most critical of all for our purposes here, there is evidence of an undeniable purchase on the life of the city, the street, and the popular imagination.

Yet fashion is trifling when compared to the explosive influence of media design. Brat pack heroes and role models today are as likely to include graphic designers and typographers as sports figures, rock stars, and film idols (the once de rigueur electric guitar, former pivot of the male adolescent household (and psyche), may already have been replaced by webmaster software packages, crossplatforms and high-speed IP connections). Graphic design, filling the void carved in the 1980s by the disappearance of both high art

First published in *ANY* 21, "How the Critic Sees" (1997).

ROMAN NOIR

and serious popular music, has resulted in a full-blown but still incompletely acknowledged, post-boomer rock 'n' roll revolution, although this time for the eyes and no longer the ears. The phenomenal proliferation of art direction in sports, music, and lifestyle magazines, youth-targeted advertising, CD packaging, film title sequences, homepages, MTV programming, lifestyle-gear logos, and related clothing and paraphernalia together form a seamless performative mesh, a cultural project in the fullest sense of the word, one of nonstop modulation and adrenalated *display*. The design heroes of the popular imagination today—and one might fix the incept date of this new regime as recently as 1992, birthtime of Aldus Corporation's Fontographer software <font merge> command,[1] of *Ray Gun* magazine and the emergence to "school" status of its niche-making R&D department, the *Emigré* collective—are those whose task it is to bestow the queasy, modern, disaffected "look" of our digitally poisoned, post-literate world: the communications systems shape-givers. Even cinema is losing ground to the invasion of the real-time total graphical interface and the video game.

Will there even be architects in thirty years? This is a question that no other generation has ever had to consider. The falsely assured still maintain that, economically, buildings quite simply are like food—irreducible facts of life that no transformation of social or market forces can ever change. They may well be wrong. Despite the customary, fashionable genuflection toward infrastructural questions and concerns today, little attention is being paid to the more radical, more disturbing reality: that infrastructural demands are not only becoming exponentially more importunate today but that these infrastructural demands *are breeding and mutating in kind and not only degree*. We have no choice today but to deal with the new "soft" infrastructures: knowledge infrastructure, program infrastructure, cultural infrastructure, virtual infrastructure. The demand for design—and de-design—in our over-engineered, over-mediated world is both enormous and pervasive, yet the majority of architects still respond to it with the medieval language of the stoic, autonomous building. Today's design world is stratified, with an emerging class structure, its associated embedded conflicts, and an emerging new proletariat increasingly separated from the principle means of production. Architects smile from behind their bow ties, for they have not yet seen that this new proletariat is themselves. And yet, surely they must know.

Our environment is being reworked in depth, no one denies this. Buildings, however, are not oases from history (unless explicitly and brutally organized to be so, and then at best only incompletely) but rather *relays* in a comprehensive cultural system of management, administration, and engineering of human affect and historical unfolding. Like the coils of an anaconda, loop after loop of the soft-infrastructural mesh is drawn daily around us, not to crush, but merely to restrict expansion in unsanctioned directions, to guide all movements subtly but uncompromisingly toward other ends. Architecture imagines itself regal, leonine and sovereign, but today it is the anaconda that rules.

How long has it been since buildings could claim—from a simple historical-materialist point of view—to represent the principle greatest repository of accumulated capital in our culture? Or more simply, when were buildings first displaced by public works projects and private industrial initiatives as the largest and most expensive things? Is not the answer—of which we all nurture our favorite chapters—the real secret history of the present, the noir narrative in which architecture figures as the delusional, vertigo-struck, and now tumbling anti-hero? The simplest answer of all, of course, is that architecture has never been the vastest, most intensive, form-giving cultural enterprise; that position has always been occupied by war. In war the soft-infrastructural enterprises at once find their model, their origin, and their reason for being. To the classic military triumvirate or C^3—Command/Communications/Control—correspond not some but perhaps all of the new design environments and milieus that have started to appear around us with such force and energy this past decade.
This is not to say that such practices have anything evil, murderous, or destructive about them, only that the citadel/fortress obsession—the well-known "edifice complex" of architecture, source of its hubristic claim to preeminence in the design world—has finally completed its inevitable, de facto evolution to a *logistics* operation. In other words, it has assumed its modern destiny as an informational regime oriented to performative environments, to protocols, and, *in extremis*, to psychological operations. This partial evolution also suggests that a hilarious spectacle is now preparing itself as our Big Building Field Marshals polish their chest medals in proud anticipation of their next Patton-like tour of what they believe to be today's architectural battlefield. If they listen, they'll hear some gunfire in the far distance, but just barely. Mostly they'll simply hear the rattling of their own medals.

***

In 1997 in Bilbao, amid the machinations of an endless political chess match, a building of undeniable beauty and exuberance rose up before a hungry if cynical world. (Nothing could have been more calculated than Guggenheim director Thomas Krens's corporate "acquisitions and mergers" gambit, or the Spanish government's tactical engagement with the powder keg of historical and geopolitical forces (an unconcealed mixture of potlatch and détente), and no one more willing to oblige these exercises in realpolitik than an architect—"I only do buildings!" he says, disingenuously.) Yet the unveiling of the dancing behemoth of Bilbao, like an electrical shimmer on the twenty-first century's terrifying horizon, has been greeted with a hysteria redolent of nothing less than a mystic visitation. I myself once believed that Frank Gehry's Walt Disney Concert Hall in Los Angeles might well become the great architectural project of the end of the twentieth century, and I was untroubled for a time to state so in just those millennial terms. Today I feel less naive. The Bilbao museum, however, and more important, the critics' and public's neurotically euphoric reception of it, seem to wish it to stand in as the fulfillment of the earlier project's lost promise. But the shortfall is simply too great. Not, of course, in relation to the earlier Disney, for Bilbao represents a significant accomplishment in its own right, though it vies for stature within a radically different context of meaning. The Basque (offshore) setting both isolates and intensifies the architectural gesture in intellectual, existential, and historical terms, but it also banalizes it significantly. This shiny new museum for the Common Market (the strident European demand for political stability, driven by the requirements of monetary interactions, need observe no boundaries as it carpetbags its way to internationally admired means of social control via architecture) simply does not qualify as a summary statement on the relations of society, form, and modernization at the end of the great modern century. Of course, many will argue that even the idea of such a summary statement is precluded by the very lessons and predicaments that characterize the thought and history of the last hundred years. That, however, depends on how one places oneself in relation to history.

It would behoove many of a certain generation to celebrate Bilbao as a concluding dance and a death rattle (a sound so eerie and unlike any other that it often leads loved ones to imagine the dying are momentarily coming back to life—whence the raptures of the mystic visitation). While Frank Lloyd Wright's

New York Guggenheim is certainly among the most radical buildings ever built in the West, Bilbao is rather at best a lyrical *son et lumiéres* show on an all-too-classical theme (though to be fair, innovation of the former kind has not been attempted often in this generation by anyone, and Gehry himself has never made claim to ambitions beyond the way buildings merely look). The pilgrimage that so many made to the museum's opening in 1997 was admirable, large-spirited, and certainly invigorating for the design profession, though deep in every pilgrim's subconscious, it also covered over an undeniably morbid, complacent and self-congratulatory act. I myself, for example, would have been as fascinated, and troubled, to have wandered the haunted chambers of the submerged *RMS Titanic.*

In the era of the ships, when architects and intellectuals were spellbound by the technical revolutions of mass travel, by the cult of speed and the plastic deformations of time, by synthetic total environments and their emancipatory possibilities for stale and weary "historical man," by the machine and its possibilities for human and collective transformation at once political, economic, aesthetic, and sexual, the *Titanic* represented an accomplishment—a polemical and aesthetic summary—of everything good, bad, and unforeseeable that modernity might become. The event of its sinking was probably the only modification that could have further intensified it as a focus of cultural longing, speculation, and fantasy. It already figured among the greatest engineering feats, technical and even social accomplishments in the popular and social imaginary. Important buildings of the era—the Fagus Works, the Monument to the Third International, later the Dymaxion House, the Empire State Building, the villas of Le Corbusier, etc.—could articulate their greatness convincingly only in humble reference to the *Titanic* and its ilk. Once long ago, ocean liners had all but stolen architecture.

As architects were preparing their valises and black ties for the Bilbao opening extravaganza, some of us, on this side of the Atlantic were deliberately heading in the opposite direction, westward to the Mojave Desert to celebrate one of the major barrier-shattering acts of the modern world. For that weekend in October also represented the fiftieth anniversary, and the commemorative re-enaction, of perhaps the second most significant accomplishment in the history of twentieth-century space. Between 1905 and 1917 the Newtonian model of rigid, absolute space fell at the hands of Einsteinian relativity and plasticism.

On the 50th anniversary of Chuck Yeager's historic flight, the Thrust SuperSonic car broke the sound barrier on land, causing the floor of the Black Rock Desert in Nevada to ripple with shock waves for 100 meters.

But on October 14, 1947, the limitations of speed and altitude (or, for those living on a round planet, cosmic *distance*) with which humans were able physically to converse and interact with their world were eliminated. General Chuck Yeager's accomplishment—flying an aircraft faster than the air could carry its sound alongside it—not only opened the supersonic realm to technological manipulation but made it possible to imagine thrusting guided objects beyond the earth's atmosphere, indeed, beyond even the influence of its gravitational field. Space had for the first time come within practical human reach. The modern world, as I have argued elsewhere, was in a peculiarly American sense reborn after World War II in relation to these events.[2] But the focus and intensity, due to the abstract nature of Mach 1 and the esotericism of its technological apparatus, did not penetrate collective consciousness in the same way as did the cult of the great liners of the early century. Had architects been less obsessed by their precious "modern movement" in the 1940s they'd surely have seized on one of the strangest, and related, experiences of modern times: the incremental destruction of London by the V2 rockets, or buzzbombs, missiles that could penetrate the urban space and wreak their destruction several seconds before they could even be heard! The World War II supersonic rockets, like the trenches and no-man's-land of the first Great War, penetrated cultural consciousness deeply, yet passed over the architectural imagination (and conscience) almost entirely.

Today we are not so easily spared, or rather we are not granted our innocence with the same casual impunity. The editor of an American design magazine, introducing a special issue on Buckminster Fuller in 1997, was compelled to described a mystic visitation of her own experience that occurred while attending Pasadena's Rose Bowl parade.[3] As the parade began, a United States Air Force stealth bomber flew overhead and buzzed the stands—"silent . . . alien . . . primitive . . ." in her words. The crowd gasped as the aircraft's implausible, inscrutable image and muted terror engraved itself indelibly in each of the spectators minds. Weeks later when recalling the event in her editorial, its power to trouble not only the mind but also the complacencies of the architectural discipline remained undiminished: "What astonishes today is aeronautical design [not architecture]" (her own son dreams of designing Boeings, not buildings). What characterizes the history of aircraft design, as well as current industrial, product, equipment, and communications design, but which can almost no longer ever be said to be true of architecture, is the astonishing flexibility and transformability of the tolerance *envelope*, that is, of the system of givens within which design takes place and whose parameters are accepted as a priori and fixed. In other words, the ethos of these practices is *experimental* in the fullest and most dangerous sense. We should therefore, in relation to the field of architecture, allow ourselves to be amused, but certainly not so easily be duped, by pseudo-innovation, by the mere variations on familiar themes of which we see so much around us today, variations that only become more extreme and strident as the envelope within which they are designed becomes increasingly brittle and stale.

As we made our way to Edwards Air Force Base we moved liked animals.[4] We neither expected, nor wished, to actually see anything (military security, in anticipation of our arrival, had in any case seen to that). We came rather to listen, to hear the sound of our century reaching back to communicate with itself, to hear the sonic boom from Yeager's F-15 roll across the high desert floor, roll across the last half-century like an auditory territorial marking, calling out to the savagely dangerous and primitive X-1 winged rocket bomb in which he first tore through the sonic wall and changed the shape and dynamics of our world. We came because we believe in shock waves, we believe them to be part of the music of modernity; not something to watch a ribbon be cut from, but something to feel with our diaphragms, eardrums, genitals, and the soles of our feet. We wanted to be in the desert badlands that day with nothing but the sun, the baked dirt, the pneumatic tremors, and the unbroken horizon. We came because out there somewhere we knew was the zero-degree and the future, and that in Bilbao was the past.

1 Acquired by Macromedia Corporation in 1994. The merge and crash capabilities were introduced in release version 3.5 (1991).
2 Sanford Kwinter, "Flying the Bullet, or When Did the Future Begin?," in *Rem Koolhaas: Conversations with Students* (New York: Princeton Architectural Press, 1996).
3 Cynthia Davidson, "Dear Reader," *ANY* 17, "Forget Fuller?" (1997).
4 I was traveling with Jesse Reiser.

To manifest one's ideas in the old way—*en masse et en bloc*—as a collective and a front, with pack loyalty, shared concerns, a willingness to appear faintly ridiculous and to defend one's cohorts in the act, is a form both of love and ritualized aggression.

It is especially hard to work as a group—but it is important, and even if, for certain historical reasons, it has become less credibly important, then it is still worth doing for at least two secondary reasons: first, because it is great fun, and second, because it goes so against the grain of our times. My own group has been mostly a failure up to this point: travel schedules, time zones, egotism, laziness, payrolls, unbearable career pressures, remoteness, etc.—all have prevented us from coordinating our efforts and thoughts in the natural and customary ways. It unfortunately even goes without saying that the following comments have never been shared with the others. But it is them whom I am addressing here—for "they" includes "anyone" who should feel him- or herself interpellated (to use that hideous old word), especially all of our known and as-yet-unknown friends and fellow travelers out there, who, like us, are looking for, and just now beginning to find, something *different*.

It is assuredly not a club that one is seeking to form, but rather a "front," a line of resistance, a point of *catastrophic* irruption. A group needs many relays, it needs diversity and dissent, and it needs rich and responsive feedback mechanisms in order to become robust, flexible, "intelligent," unpredictable and alive. To those who dream of collaborative endeavor let this ANY conference serve as our "host" in the biological sense of the term; let us have a working session, let us publicly acknowledge our commonality, whether it is real or only willed (and all the more open for that).

The comments I have assembled here are but little sparks, provocations meant to start fires in the best-kindled camps. They are brief. If what follows is a text, then it is an incomplete one, an *aborted manifesto*. When I first began thinking about it, I called it "A New Organon" (after Aristotle, Bacon and Brecht). Manifestos, after all, even aborted ones, need to be about a "new" something.

It no longer seems to matter—not essentially anyhow—*what* one does. The infernal genius of the market is willing to cut a convincing swathe for anything, however hollow. What does matter though, more than ever, is *how* one does it.

LEAP IN THE VOID: A NEW ORGANON?

This is where a "subtle coup" is taking place in the architectural field. The practitioners I might cite as examples would not themselves immediately recognize their practices in the following descriptions. Yet once they have done so, they will realize that they have already abandoned architecture in the traditional sense, and once so emancipated, they will not willingly be going back. Architecture is already no longer the familiar cult of objects. It is becoming an organon, that is, a system of investigation, invention and technique.

We—whomever in the end we turn out to be—are no longer interested in what the last two decades of pedagogy and advanced practice has referred to as "process." For process has become little more than a poor man's Game of Life,[1] a parlor game for those short on intuition, a ruse with which to cloak the deep fear of matter that has characterized architectural enterprise since its mathematization and routinization by medieval template stonecutting and Rennaissance perspectivalism. The ideology of process—often a desultory fishing expedition for a formal parti based in remotely related, but more rigorously organized forms of knowledge—is a postmodern crypto-transcendentalism that does little more than impose arbitrary routines onto the logic of formation. For a period of time, design process ideology, in schools and in so-called advanced offices, permitted a necessary opening of architectural practice away from the parochialisms of aesthetics, styles and manners. It permitted, during the postwar pseudo-avant-garde era, a direct engagement of architectural discipline by the most sophisticated concerns of cultural and intellectual practice. But this engagement was at best clumsy, graft-like, adversarial, at once forced and ponderous. This is because process almost always meant "automatic process," for it was naively believed that removing the human factor from the design loop would both free the resultant work from regressive forms of subjectivity while preserving all the potential and the rich endowment of the mind. The partisans of process were wrong on both counts. The ensuing stylistic cults of complication did little to hide from the historical record the fact that what one had to do with here was just a new logico-mechanical reductionism. The new poverty was mistaken for radical anti-humanism. In America, its remnants can be found everywhere, and no less so amid our own ranks.

This talk was presented as the opening lecture at the Rotterdam ANY conference, June 1997.

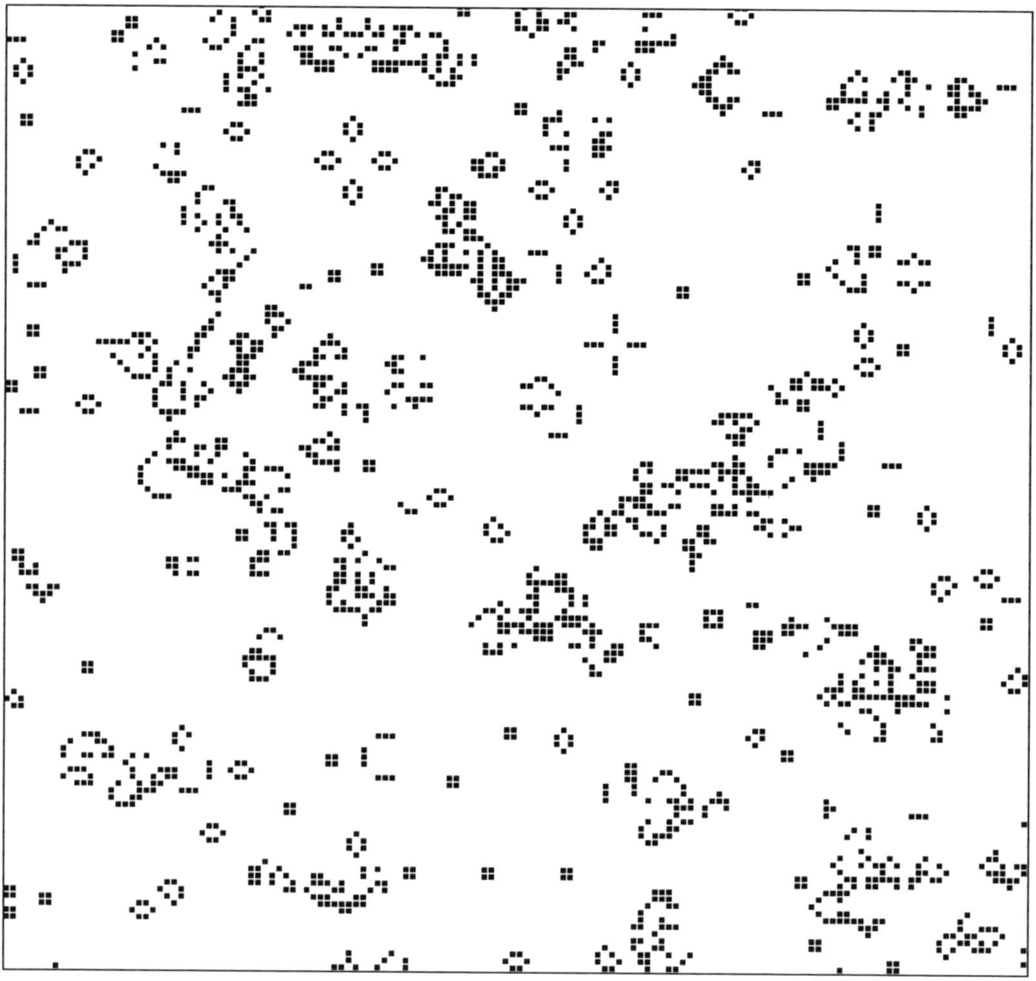

Game of Life, John Conway, 1970;
where the cellular automaton was first introduced into our unsuspecting ecology.

Today, if one looks at it from a particular perspective, architecture has begun to vanish as a discipline, and some of us are not mourning. More and more, we like to think of practice in far more generic and elastic terms, we think of what we do as *design*, and like the generations before us, we feel the need for an escape velocity that might carry us beyond the sclerosis of inherited boundaries. For us the new design envelope is an organon-in-the-making, it comprises a will to technique and an ethos of research in real domains.

What are technique and research in real domains, and how are they being used to achieve an escape velocity sufficient to free us of the gravitational field of "mere," or reductionist, architecture? Technique, as we understand it quite simply, is the authentic and *immanent* version of what 1980s and 1990s process ideology could only crudely simulate. Technique is the engagement of real logics present in the human or non-human environment and their conversion into potential—specifically, into apprehensible, *formative* potential. Technique is design from within.

Real research represents a more classical formulation. By it we would understand the idea that investigation is not simply a *part* of, a preparation or inducement for, design practice, rather that design practice is fully coextensive with investigation and the production of knowledge and correlative design culture. In the case of technique, the logics are understood to be there as a kind of immanence or pregnance in matter, and they are followed in matter because the world is only matter and its products and nothing besides. For this reason we have referred and will continue to refer to ourselves as materialists.
But real research adds a dimension of ecumenism: it claims that the entire material world is the product of shaping logics, and that these are in turn the proper subject and domain of design research. To follow the grain of the world and to see in this infinite twisting dance the expression of multiple, converging formal and diagrammatic logics, are both the a priori and the mandate of design research. As materialists, we do not approach linguistics or economics to extract from them a method or to pillage their banks of metaphors; we see them as themselves material expressions of exogenous (external) and endogenous (internal) shaping forces.

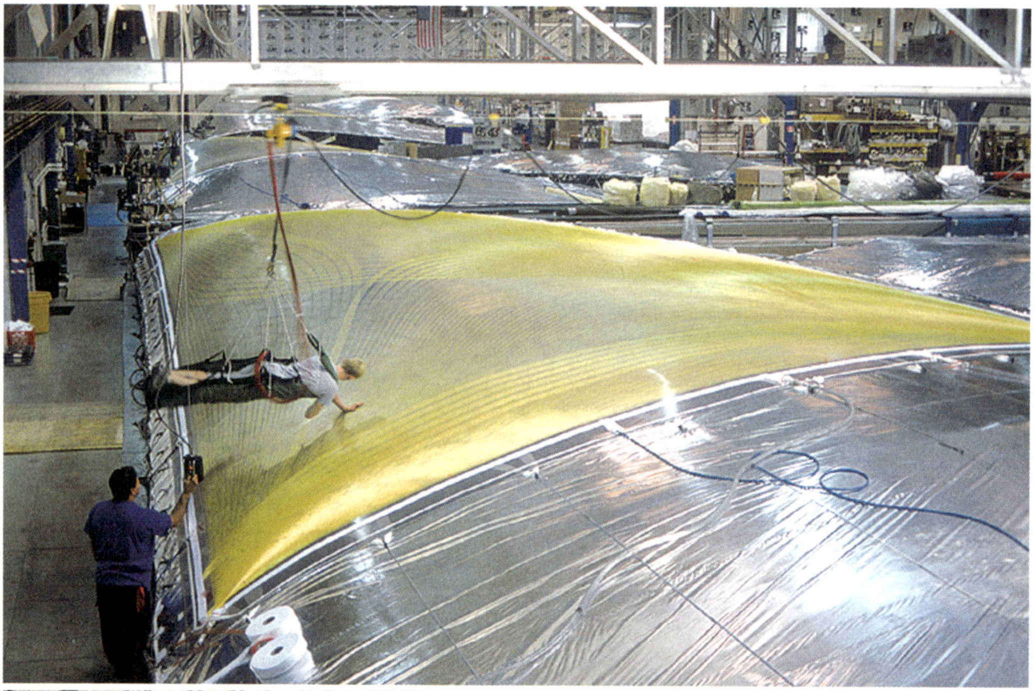

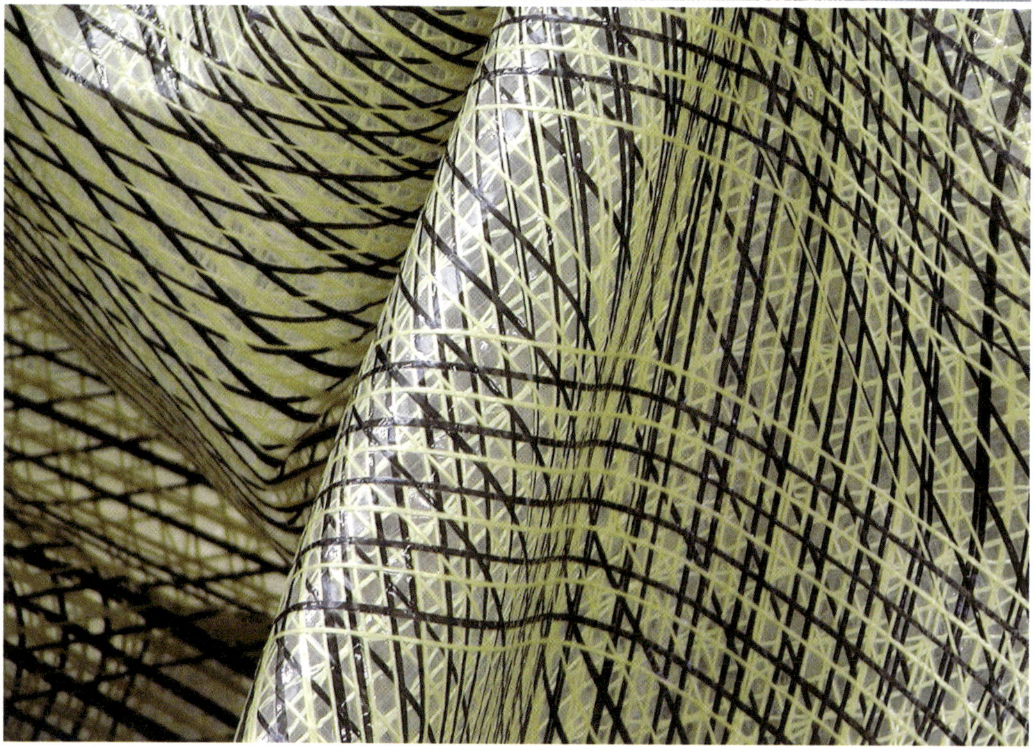

3DL Sail. A molded carbon/aramid fibre composite material shaped into an essentially solid three-dimensional structure based on anticipated load paths. All that is flexible becomes solid and all solids become flexible as engineering invades the last margins of exploitable advantage in the material world.

We do not believe in models and we do not seek to "apply" anything external to architectural or design practice. Rather, we extend the domain of design possibility by following research to its inevitable, yet unpredeterminable conclusion. We produce works of incredible ugliness, novelty, elegance and naivety. Yet such adjectives apply only provisionally, for these are not works of architecture, *but experiments in design logic, research and potential.* Some of us are engaged in what might be called teratological practices, others in extreme hybridization and cosmopolitanization. Either way, we all seem to specialize in what might be understood as monstrous evolutionary deviations from common or standard types. The crux here may be found in the word *evolutionary*; it makes clear claim that the logic of development forms the substrate of design action and intervention, not arbitrary production *ex nihilo*. The immanentist credo holds fast: design accelerates, channels, or modulates, flow; and it does so as an embrace, as a flow itself.

The ultrahybridity, the teratological, in relation to a norm, is commonly said to present a *pathological* phenomenon. Yet, as the great neurologist Kurt Goldstein once said there is substantially "greater revelation" of a living logic in pathological, rather than "normal," phenomena. He argued that it is a mistake in both method and understanding to see in the altered or impaired organism a merely reduced state; rather, we must see in the whole constellation of disease's bizarre behaviors and morphologies acute and creative expressions of a robust will to actualization that actually *builds* innovative behavioral pathways around the damaged parts. Herein lies the key to a successful escape velocity: let research follow the real, let it be encumbered by no moral and aesthetic preconception, and design will follow as an integrated process.

The late twentieth century may one day be known as the dawn of the age of the algorithm. If so, we wish to be the first to embrace the new rationality that sees space and matter as indistinguishable, as active mediums shaped by both embedded and remote events and the patterns they form. Ours is a transactional world, not a deterministic one, though we do not deny the efficacy, the economy and the beauty of the mechanist, and ultimately classical methods that have brought us to where we are. Yet, architecture is among the only arts that has not undergone an anti-classical revolution in this century. After all, even late-modern fragmentationist ontologies were deterministic.

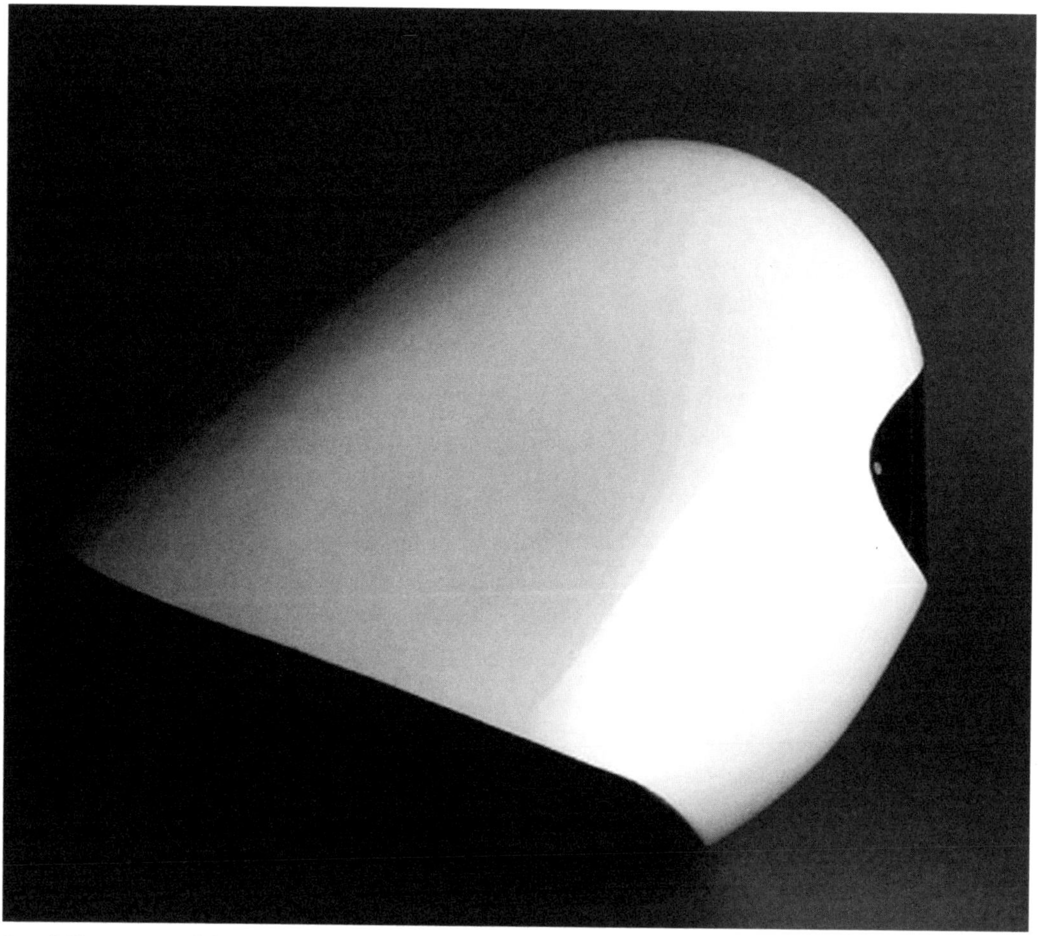

Speed Monster speed skiing helmet. A sustained essay in shape grammar from its atomic structure to the mathematical contours of its performance envelope: a near-perfect virtual decapitator. Helmet design by Composite Beat Engel.

Now, however, seemingly indifferent to the pressures of the *fin de siècle*, that revolution is taking place in spite of itself, silently and ineluctably. Henri Bergson referred to reality as an "intensive manifold," a matrix of interpenetrations whose fundamental characteristics are tension and necessity. As the first philosopher of evolution, it was his task to explain just how it is that matter came to be fraught with necessity, how indetermination came to inhabit the actual compounds that make up our world. We too believe in this necessity of matter, its impatience to become. At root it may even be construed as a political concept: Evolution is nothing but the gradual insertion of more and more freedom into matter. In that sense, we humans, custodians of the most advanced form of Mind, are simply the most free material entities in the universe. But the universe itself is not all that far behind. It too pullulates with openness and indeterminacy, different from us, though perhaps only in degree.

The computational paradigm that is overtaking us—and this paradigm has nothing to do with computers—has extended our concept of materialism to regions of reality—that is, of society and economy as well as of nature, which paradoxically are often fully incorporeal, though no less replete for that with agency, with tension and viscosity, with necessity to organize, to change and to move. The shapes of our world, those that determine our being, are today more than ever invisible, embedded: they are the shapes of time.

Matter has always given special importance to transitions, to thresholds and discontinuities, to pregnances, and to differences (to the delta T, or changes over time) because this is where all "potential" to change or to become is lodged.

By bringing the seed of "form" and "design" to these wet sites of spontaneously unfolding potential, we believe it is possible to offer shapes and diagrams to the intuition that go beyond the three-dimensionality of the mechanistic worldview, indeed that allow a more or less faithful "reading out" of the "intensive manifold," of the increasingly obscured dynamo of interpenetrating forces that drives our social, economic, technological, and natural worlds today.

If such a program were somehow to be construed as merely a new form of mimesis and we as "expressionists," then I myself am happy to be one.

1 John Horton Conway's computer game (1970).

Ceci tuera cela. Buckminster Fuller rethinking weaponized infrastructure.

Mainstream architects—with the notable exception of Louis Kahn—have never recognized the significance of Buckminster Fuller. Yet the more appropriate question today is, *do they deserve him*? No other discipline has survived from the days of the Trivium and Quadrivium with as small-minded an outlook as architecture. Brunelleschi's dome, for example, remains acceptably part of the aesthetico-ecclesiastical order that current "avant-gardists" continue to worship; Fuller's languishes. No other discipline makes small-mindedness a virtue quite the way architecture does, and nowhere as dogmatically as in its knowing denunciation of overextended ambition, unbridled invention, systematic theorization, obsessive technical rigor, sweeping universalism and anti-classicism—in a word, of Fuller. One might say, in the tradition of the Italian Futurists, that Fuller was the first truly *secular* architect, the first to embrace synthetic ideas and the brute realities of modernization to the total eschewal of tradition, its glories and superstitions. Fuller was the veritable Einstein, the Schönberg of architecture, but architecture was never interested in real revolutions. Its business, it believed, was gravity, and it was comfortably, though cravenly, ensconced within the fey yet impregnable citadel of art history. Fuller revealed "modern" architecture to be an imposter, and in a treacherous, effective summoning of its own army of clerks, the architecture establishment succeeded in exiling him to its lunatic fringe. But Fuller was indomitable because he was the first to truly marry design to science and philosophy, and no one could then, nor now, dispute the following point: he alone was responsible for more fundamental innovations and transformations of thought than the entire profession could muster or make claim to in several hundred years.

For decades one was well instructed not to mention Fuller in architectural circles, lest one bring down upon oneself the wrath of the clerks. Today, however, it has become strangely possible to speak to the subject. What has changed? Perhaps no more than the following fact: Architecture, moribund for so long yet fatally vain, has only of late had the clear, sobering glimpse of its own mortality. The elements contributing to this perception of mortality have been rising, like the formations of an attacking army, all around it for a century, but the central fact has been driven home by a single point of pressure: the economico-demographic dynamo of ultra-rapid, hyper-rational *development*. The era of naive infill, of the isolated building or building system set in a firmament of immobile forces, is over.

FULLER THEMSELVES

The brave new world we now face—and there is no looking away—is constituted by the media-economic imperative to produce "urban fabric" everywhere, even where no concrete or historical firmament for it preexists, where all is a-swirl with abstract "infra" structures waiting to be captured, concretized, set into play. But designing buildings and designing *development* (routines, propensities, unfoldings) could not be more different things. Development poses the unprecedented challenge of giving form within the agar of politics, history, media, demographics, knowledge systems and economics. In a sea of dammed but invisible forces, it is an art of total deployment: full-scale, all out, comprehensive systems *design*. Architecture could never figure as anything more than an incidental relay before the brute fact of a society's *total technical deployment*, the synchronized mobilization of all of its machines and its tools, of its physical *and* intellectual resources and life. This is what once made Fuller intolerable to all architects *bien pensants*, and of course, what also makes him so necessary for all of us today. Futurism formulated it, Moholy-Nagy probed it and began to repertory it, but only Fuller, in his hubris and madness, systematized it. Design, he showed us, must attack the entire world; its task is to produce "advantage" over adversity or hazard, to embrace work-potential wherever and whenever it exists. Most work-potential in advanced societies is lodged within patterns of knowledge and patterns of matter and their fluid interplay: that is, in shaped action and configuration. Every pattern, Fuller understood, is something *made*, it is a modulus of trapped potential (shape), a system or reservoir of active storage and retrieval.

Fuller was the first designer in history to understand structure as a pattern comprised entirely of energy and information. In this, his precursors are more Carnot, Clerk-Maxwell or Boltzmann than Viollet-le-Duc or the masters of the Gothic cathedrals. Fuller never discussed design uniquely in terms of buildings. He conceived of the universe itself as an energetico-informational continuum, something dynamic, and always transforming. Fuller was the first designer to think of form in terms of consequent *phases* of forces and matter, and though he did not apply his attention to the dynamics of its temporal partitioning (i.e., its productive *dis*-symmetries), he may certainly be said to have anticipated by decades the morphodynamic insights of mathematician René Thom. Energy, he surmised, was conserved. It was saved in *form*, that is, saved informationally in pattern-work.

First published in *ANY* 17, "Forget Fuller?" (1997).

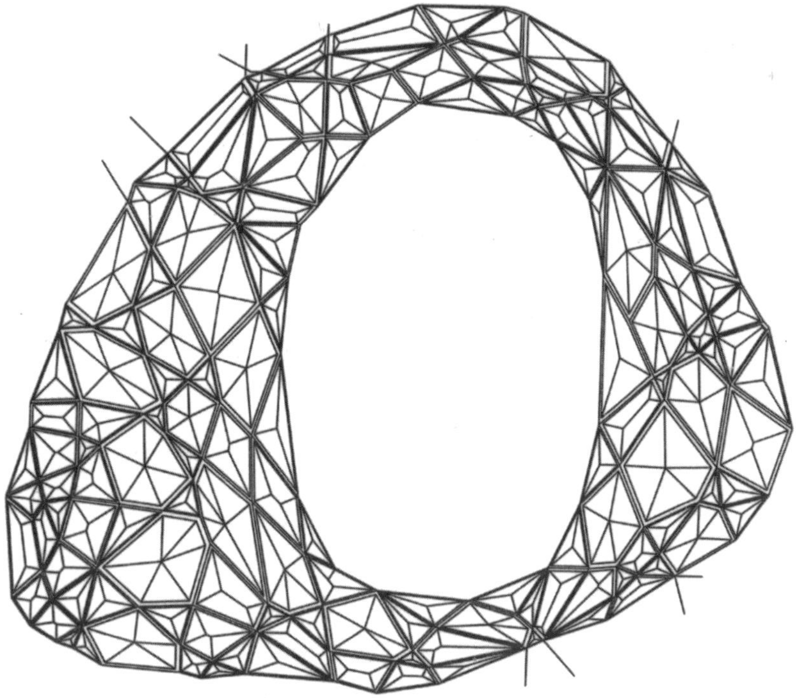

THE NEW IMMANENCE

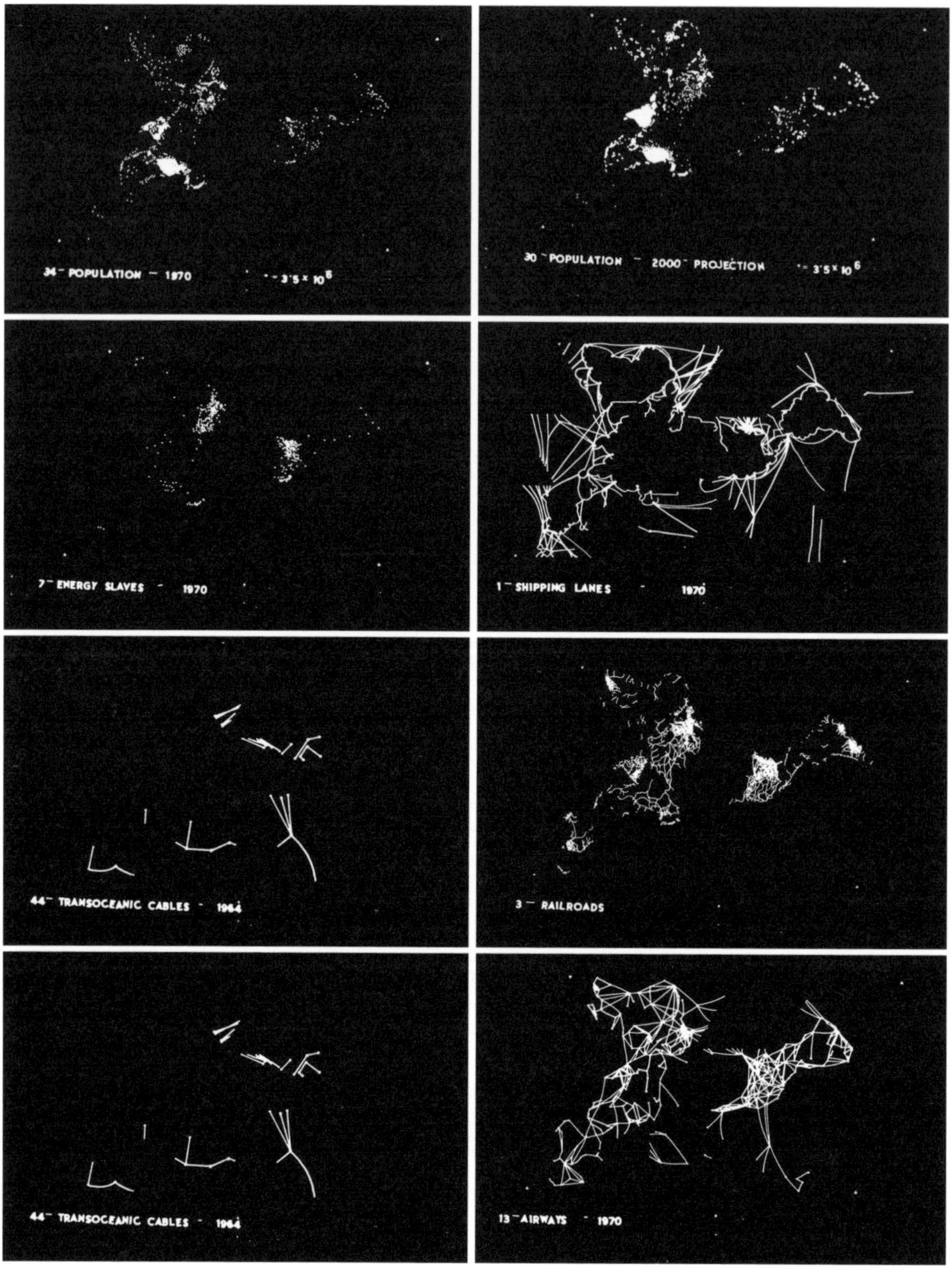

World Game, Buckminster Fuller. A global resource simulation display designed as an alternative to —or sober updating of—the geopolitical war game.

Though it was a fixation with him constantly to assign strict inside and outside perspectives in his inventions, the principle of global fluidity was always maintained through ritornello concepts like succession, events, regeneration, modular frequency, geodesics, and the irrepressible patterning processes created by the inherent need for charges (loads) to flow from one place to another in the universe. Fuller often returned to the concept of the "alloy," the chemical or metallurgical mixture whose properties are unforeseeable and irreducible to those discoverable in the metals' separate states, yet which in amalgamation reveal new efficiencies and form a new type and order of machine. Here again, capacity for *action* and expression, and therefore new types of economy, is always the end product sought for by design philosophy and design science.

Few of Fuller's critics have understood this. Fullerian science seeks to isolate not only visible patterns but informational, "cybernated" ones that express themselves in disciplined theatrico-mathematical performances. His championing of the principle of tension is one complex example: new molecular arrangements (materials) make possible new physical-mathematical ones (models), which in turn give rise to new aesthetic-material ones (built structures). All of this, however, remains concretely dependent on broader schema-patterns such as social relations and industrial production routines. He saw the world economic system, earlier and more explicitly than anyone else, as itself a mega-alloy of gigantic proportions and complexity, sweating properties left and right, a geometric modulus of existing and possible performances determined only by the particular phase in which it is at any moment caught. He saw the total infrastructural Gestalt that underlay every technical performance or design actualization. He knew the alchemy of interstreaming numbers—financial, chemical, industrial, physical—just as certain materials scientists are said intuitively to "feel" or anticipate the interaction of molecules in composites. His understanding of the nature and importance of information led to the first call for an "anticipatory" design practice. (Today one might speak of a predictive design, in the sense in which some scientists at the Santa Fe Institute for the Study of Complexity have applied similar concepts to financial markets.)

Numbers may well be abstractions, as mathematicians and philosophers assure us, but their interactions within moving streams—equations, algorithms, and continuums of matter itself—are manifestly real. Yet numbers in motion interact in both predictable and mysterious ways, like the cross-reactions and webs of organic chemistry, leaving ample room for both certainty and intuition. A flow of numbers, for example, can act as an acid or a fixative, or indeed as a catalyst to generate or transform another numerical set, giving place to a world of objects, actors, forces or effects that cohere, give meaning to one another and provide at once reassuring stability and a guarantee of steady change.

Cecil Balmond is an engineer who has seen deeply into the reality of numbers as living agents of physical organization. His approach to them is intuitional, which at bottom means *mediated by form*. Like certain gifted athletes, Balmond manipulates numbers geometrically: he principally draws lines, matches or connects one pattern to another, and creates meta-patterns which produce "solutions" in two senses of the word. Initially, the solution to an engineering problem of statics: the solving of a variable that satisfies an equation to resolve the way forces will be made to flow through a material to assure the stability of a structure. But solution as well in a most unconventional sense, one that could be said to represent his signature style: Balmond's solutions always represent the "dissolution"—and therefore dissimulation—of structure within form, in just the way a solid (salt crystals) can be said to be dissolved within a liquid (water). The concept of a column, for example—you will see none in Balmond's buildings—is anathema to him, the sign of a failure to channel forces efficiently into and across the flow of form itself. Balmond is the designer who invented the Archimedean move of amplifying a cross-braced truss to such colossal dimension that it no longer serves a subordinate role (as an integer merely supporting another, more principal one, such as a slab of roof or floor), but is now inhabitable itself, environmental in scope, and invisible, because so integrated into the large scale pattern of the building. Engineering thus becomes totally immanent to the materials and the geometries that it inhabits and guides.

Numerological mysticism? There is in Balmond no doubt a touch of this. Yet since the demise of mechanical explanation in science at the end of the nineteenth century, problems of asymmetry of all types of structure have surged to the fore. This was a moment in the history of knowledge when Romanticism (Goethe's and Coleridge's) offered an objective advance to classical material thought. Today, Balmond's defiance of the platitudes of the Newtonian engineer have opened up a world of demonstrable realities—his word is "informality"—that call for not only a new type of knowledge but a new ethos of design, one that obliterates the petty functionary's distinction between engineering and invention.

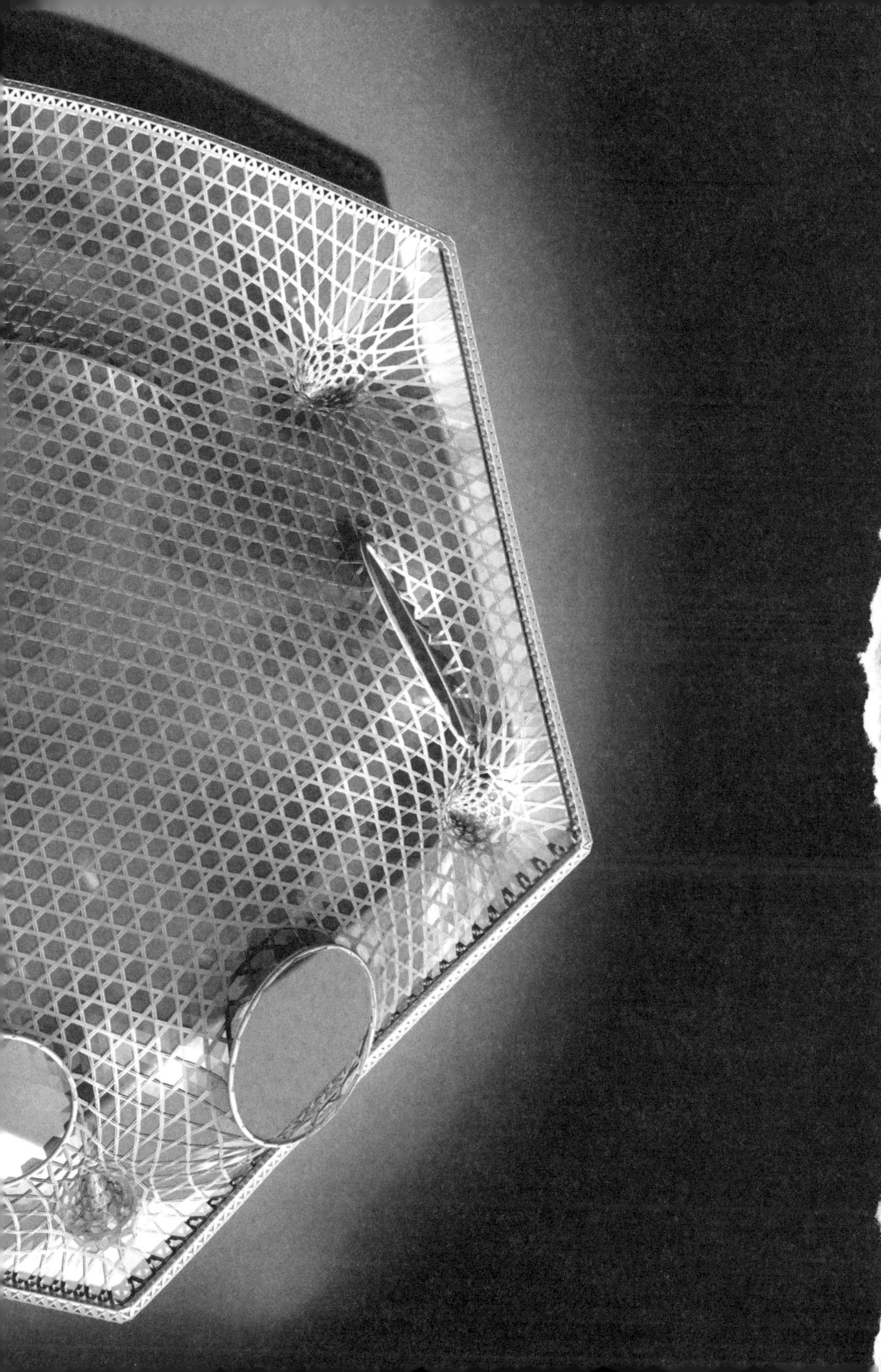

Fuller was the first to understand the possibilities and critical role that telephony, telegraphy and radio could (and indeed already did) play in world-informational tensegrity imaging, how "change over time" algorithms in markets, populations, and resources could be made visible and concrete and made to enter directly into design intuition. Indeed, he was the first thinker since Bergson to theorize intuition (macro-micro oscillation information processing) in scientific terms. Fuller may have also been the first person to use the word *ecology* in its current expanded sense as the patterned interplay and mutual determination of organism and environment. He was the first designer to have comprehensively understood the social and natural environment as a fluid, shapeable, and active medium. Fuller's was the first systematic "organicist" vision in the design world, the word used here not in the metaphorical sense but in the technical one that issued from biological and philosophical circles in the 1920s and 1930s around the principle of "organization." Fuller had a profoundly scientific intuition, which meant that he was primarily interested in values such as beauty, elegance, and economy as they pertained to a solution, not to cloying ornamental or stylistic properties. This is precisely the type of beauty to which "ersatz" (i.e., other) modernists had come to be attracted and which they sought to imitate rhetorically in their work. Few, however, had the ingenuousness, the rigor, or the naiveté to go all the way. The Fuller project still holds up as the far limit of *extreme* design.

Fuller understood, perhaps better than any other designer, the full importance of machines and their evolution. Increasing refinement of engineering and machining tolerances and increasing refinement in manufacturing and processing techniques allowed "intelligence" or cybernetic capacity to become more and more generalized throughout the living environment. At the level of agency, the distinction between matter, design, machine, system, routine, civilization, and universe was collapsing. But if engineering for the advanced air-ocean world (mobile or non-gravity–based structures) increased the general quotient of distributed intelligence, engineering for the primitive gravity world (architecture and its prehistoric cult of "compression" tectonics) remained a force of active stupidification. Even today the most finely crafted buildings rarely even begin to approach the refinement of facture or level of interrelatedness of a Ford Model T, let alone a nuclear submarine or supersonic aircraft. Fuller's simple point is that a large part of a society's fixed capital and public goods comprises infrastructural possibilities—techniques—that remain woefully unexplored. Here, and nowhere else, lies the true problem of design for the twenty-first century: the marriage of fixed and intellectual capital. And to get there, we must pass through Fuller.

The Fuller program has simply been too capacious for the parochial architectural imagination. Even in the closing hours of the twentieth century, at the spearhead of what looks to be a comprehensive revaluation of the multimodal Fullerian project, contemporary designers and critics still seek to do no better than to reduce Fuller's *Gesamt* design-science and utopian social project to a dismissable artifact of popular culture.[1] Fuller, on the contrary, represents nothing less than *a machine*, an arrangement of forces, of an entirely new type—a total commando urbanist, a universal software wizard (predating even Alan Turing) who literally reinvented the concept of "agency" or "program" in design practice: what it was and what it could do. Though no one ever subscribed to Fuller's naive politics (they were hardly sillier than Le Corbusier's), he often spoke about revolution and was clearly the only designer in history (with the possible exception of a Henry Ford or a Linus Pauling) who had the creative scope to blueprint the Russian Constructivists' dreams. Fuller's very existence menaced every complacent idea of architecture in the twentieth century. Of what other designer can even the following *type* of statement be said: *He saw the automobile industry as a comprehensive challenge to the sociology of design*? The capacity, not only to discern an emerging pattern in economic and productive organization, but to see it in its relation to the modern bathroom, the fluid atmosphere, and the distribution of energy in the earth's crust, indeed to see all of these as an interrelated manifold to be shaped and tapped through design intervention, is not rare, it is unique. For those of us who wish to see ecumenical design discipline succeed architecture through a new wave of tough-minded cosmopolitanism, to allow insular architects to become the historical relics that Reyner Banham predicted they would, to give birth to an entirely new type of urbanist intellectual and practitioner for us, Fuller is a beacon.

1 See *ANY* 17, "Forget Fuller?" (1997).

World record group skydive with 300 jumpers in formation, 2003.

Rarely has architectural production been so well and richly served by theory, nor theory so well-legitimated by, and so happily submissive to, production as today. It is not rare these days for the theorist to design or for designers to contribute lucidly to culture with concepts, arguments and texts. The two sectors, often separate, are so busily ministering to one another that they seem unaware of the deeper deprivations to which they are giving place. For today's lovefest hides a sad, possibly ugly side that has become a matter of vigorous indifference to everyone, simply because we are nearly all party to its interests. Just as certain economists, with clear reason, warn that current rates of growth are unsustainable, that public assets are overvalued, and that gravity continues to guide ballistic trajectories in markets and spreadsheets, yet are scorned and ignored with apparent impunity, so too has the architectural discipline all but abandoned the external and long-range viewpoints to which both its future and its soul, as well as its deepest ideas, belong. Everyone is reveling at the party, but there are no designated drivers to be found.

These are not, of course, sober times given to sober claims, and as a writer I have been no exception. I have served as an apologist for flawed architectures and as a less than temperate rehabilitator of others' homeless and inchoate concepts. Never, though, have I been an apologist for that particular type of neoconservative fundamentalism that, since the mid-1990s, has swept corporate, cultural, and academic milieus, and which is now settling into our own field like a dry rot out of control. The current intoxication with the magical flame of neocapitalism—expressed primarily through the exaltation of market forces and their legendary, but not proven efficacy and intelligence—has so subsumed contemporary culture, society, and media that even our intellectuals, if one can still call them that, can no longer muster the awareness that something troubling has happened. And yet, the discipline of architecture, once a poor relation to the humanities and the social-scientific disciplines, has today reached a level of popular sophistication exceeded only by business management theory and software research and development. But all this really means is that these fields have come to resemble one another in both substance and expression to a disturbingly high degree.

LA TRAHISON DES CLERCS

(and other Travesties of the Modern)

It is well known that architects have rarely bothered to hide their status as whores to patronage (shamelessness, after all, is the best way to conceal the humiliations of subservience, precisely because it routinizes them), but a new (not different) ethos is attempting to generalize this posture into a radical, and for the first time systematic program. Rising up like an improbable tidal wave from the dry side of the Lowlands' dikes, a new movement—Dutch this time, like the Italian, Japanese, Spanish, and Swiss ones that preceded it—is poised to seduce and submerge us with a quirky juxtaposition of toy forms and unsentimental bureaucrat logic. To date, the evidence submitted is that of a titillatingly inventive body of work that brims with wit and charm and especially paradox, while remaining—how should one phrase it?—*conspicuously* intelligent (though this depends considerably on how far one is willing to stretch "intelligent"). What makes the work topical, necessary, and immediate, according to the compelling promotional arguments that have risen up to meet it,[1] is the immensely refined and up-to-date "pragmatism" it is claimed to demonstrate, a feature that makes it, along with that of such "masters" as Herzog & de Meuron, perhaps the first really "new" architecture of the neo-modern period. Disencumbered of all social or physiological intensity—that is, of anything with the historical, existential thickness of ideals, dreams, or "transvaluing" values—the work actually aspires to nothing beyond pure "megamachine" architecture (to use the once ominous, but now denatured phrase of Lewis Mumford created to denote the vast technical knowledge systems and devices that make up modern rationalized society). Clever and playful, and never less—nor more—than briefly endearing (a function no doubt of the work's pointed renunciation of any claim to importance), this work nevertheless represents an architecture of resignation and disorientation masquerading as enlightenment, an architecture of spiritual exhaustion presenting itself as scientific activity, as hip unsentimentality, and even as a "coolness" in the technical sense of possessing an apparent poise or collectedness that reigns over a fast play of relationships. There is simply no cool, however, without magically woven *restraint*, that is, no cool without a hot substrate held in tense equilibrium.

First published in *ANY* 24, "Design After Mies" (1998), an issue dedicated to the Illinois Institute of Technology design competition and the philosophical and stylistic problems of designing a new building for the Mies van der Rohe campus at the turn of the millennium.

NAS
NASDAQ LISTED
+3.1%
19.04
+0.58
+3.3%
24.00
+0.76
+1.5%
31.54
+0.47
DENTSPLY
+2.2%
30.45
+0.64
APPLIED MATERIALS
AMAT
+2.1%
17.84
+0.36
30.88
Columbia Sportswear Company
COLM
MARKET FORCES
OPLINK
OPLK
BUY ORDERS
SELL ORDERS
+7.8%
+4.54
18.99
+5.6%
+1.00
MARKET VELOCITY
62.96
ORACLE
+1.1%
17.16
+0.18
Microsoft
+1.4%
30.88
+0.43
Google
+1.5%
495.18
+7.09
bea
+0.3%
12.41
+0.04
+1.5%
23.53
+0.34
NASDAQ
NASDAQ

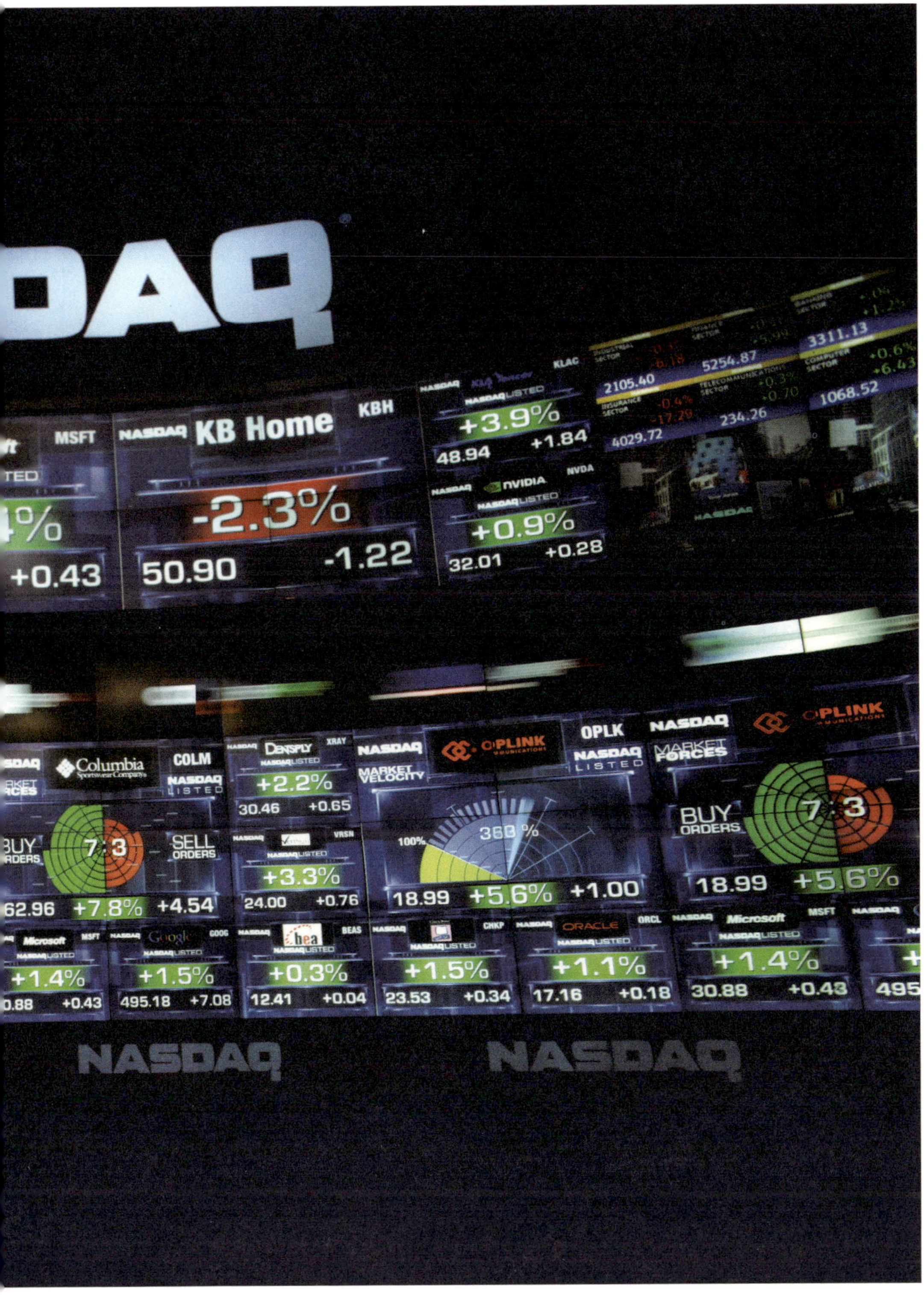

Datascape: NASDAQ ticker.

Paid Parking, NL Architects (from Big Soft Orange catalog, 1998). Car park as branded content.

The central thesis of this new school is supposed to consist in the embrace of (global) infrastructure as a shaping force, but the host of Koolhaasian concepts of which this new work is assembled, is being considerably watered down and one-dimensionalized (albeit often elegantly distilled into slogans such as MVRDV's term "datascapes") so that it no longer represents anything thicker than the rote numeric sequencing of market behaviors and demographic pressures. A naive determinism is inevitably allowed to intervene here, one that neglects the panoply of real, unstable social processes and their own infinitely dense, and ever-shifting, psychic, political, erotic and anthropological infrastructures. The "pragmatism" that the new Dutch work represents, as rich in possibility as it is poor in its current expression, is a pragmatism of the worst kind: it remains little more than ill-digested and reductionist Koolhaasianism, to which is added the bureaucrat's compulsion to justify impotence ("planning is impossible, the market rules!") by inflating his or her own ineffectuality into an historical and aesthetic ideal. In the hands of the new Dutch architects the dramatically new perception of embedded (abstract, invisible) expert systems (human and social engineering) as the constitutive shaping forces of our time becomes a glib and often merely touristic drive through an all-too-familiar media landscape of *thinfrastructure*.

Were this group of architects but a second-rate bunch, it would mean no harm done and no attention would need be paid. But the new Dutch work is nothing if not compelling and cogent, and were it not so unabashedly derivative and middlebrow, it might almost credibly stand as the "new" platform that it is touted to be. Yet because emulation plays such an enormous role in the dissemination of architectural ideas, and because so little intelligence or understanding is required simply to emulate, this work and the attenuated ideas that drive it are not only gaining wide attention but in some cases are already earning a vigorous following. That the material our colleagues and students are now perusing is not the real or even living thing but rather a more or less earnest and denatured copy of a richer, more demanding, more ambiguous, and complex original is becoming especially important now that critical unanimity has begun to join the herd in embracing facility and accessibility to the complete exclusion of difficulty and of historical density and singularity. It is not just the crushing dominance of Koolhaas's Office for Metropolitan Architecture on the Dutch intellectual scene that diminishes much of the new work.

Dutch Pavilion at the World EXPO 2000 in Hannover, MVRDV, 2000.
The new genius successfully combines tragedy with farce.

It is the imitative poverty of these practices—partially occluded by their own savage brilliance—that threatens to diminish eveyone's capacity to read and to think the more raw and resistant, the more ambiguous and unstable becomings that make up historical process and material life.

The avowed task of the new B-movie architectures is to reflect existing conditions and existing knowledge (gleaned from financial pages, official briefs and "research") and they do this well enough, but unlike the model from which they were spawned, they lack the complexity of vision and imagination and the existential gravity to project latent or emissary structures *forward* into a truly unknown future that may well be entirely *different* from the present. More than anything, what these new, lite, hyperstylists represent is the gradual replacement of *qualitative*—the delirious in New York—by *quantitative*—Maastricht bean-counting. How the cloying *Heimlichkeit* of the MVRDV universe, for example, ever cleared the critical gatekeepers in Germany and America (indeed, how it even graduated from the local Dutch joke mill) is a mystery beyond the powers of this writer to explain. There seems to be a willingness, first on the part of the Dutch, then on that of the rest of the world, when faced with the new OMA sublime (I use this word only with irony and to note the irrational fear that their activities appear to inspire in their colleagues) to take neurotic refuge in masochism and burlesque.[2]

The general silliness of the MVRDV project (as distinct from the earnest vacuousness of the over-handled Orange group[3]) is purely symptomatic, though even this could easily be forgiven were not their worldview so narrow, their gags so provincial and devoid of humor and mirth. All too happy to publicly play second fiddle, cuckold, and stable lackey, this enormous house of talent has recently committed its resources to producing a quasi-book (*FARMAX*), though only to make an embarrassing parade of their own intellectual and stylistic deficiencies, and as part of a perverse genuflection to another model (*S,M,L,XL*) that neither needs, nor frankly deserves, the apotheosis that MVRDV cannot help themselves from handing it. So subjugated by the greater sovereignty from which they have drawn, and so apparently resentful of it, their gestures of pseudo-homage are only half-heartedly covered over by original virtuosity and are simply pushed to the point of half-baked parody and plagiarism. It is hard to tell where all the humor went awry, but for the most part we spectators are left simply to avert our gaze in discomfort. Some of us.

For there is no doubt that this work has found a broad following in recent years (though much of this may be simple curiosity of the morbid variety), and there is no doubt that we have been somewhat impoverished for it. While the American school today collapses under the weight of the naive formalism and technocratism that crippled it from the beginning, a natural tendency has been building up in the system, a need to return to the realities, especially the abstract, intellectual ones, of social and historical existence. I myself remain deeply sympathetic to this tendency, and deeply drawn toward a therapeutic, antidotal passage through neomodernism. But neomodernism need not simply mean neocapitalism, nor do we need, as most of the newest and most visible design work in Europe is doing, simply to celebrate the subtle but sweeping triumph of vulgar neoliberalism.

The social imagination has all but vanished from advanced internationalist design, and this has had the temporary effect of rendering its occasional appearances as dismissably quaint, regionalist, or nostalgic, at any rate, as lacking the hardheadedness necessary to fraternize in the Esperanto world of e-commerce and fiscal globalism. The fierce inventiveness and productivity that we are experiencing in so many of our social and cultural economies today is obscuring the emergence and spread of a new poverty, the poverty of *liberal* pragmatism.

We designers are the inventors of images of freedom. Yet what has become corrupt of late are the means by which the last term is defined: no longer the freedom to transform absolutely the very possibility of what might come, but the freedom merely to harvest, in relative fashion, temporary advantage from the world that already exists. This does not mean our world and our culture are any more fixed or stagnant than before, only that their course or development has become no longer even theoretically subject to design intervention. This is the failure of *our* practice and theory and no one else's. The new ideas we believe to be celebrating in the new European work, and in stylistic neomodernism in general, are but half-ideas, recycled clichés and fiscal propaganda (consider, for example, how powerfully and smoothly they have been integrated into the British shopping scene and retail culture by *Wallpaper** magazine). As long as markets continue to soar, the neoliberal logic that is being used to excuse (though not explain) them will continue to inspire insipid, soulless, and complacent "superstructural" works, because the measure of what permits emancipation from mediocrity in this and all other times does not belong to the world of the balance sheet, but to that river of indeterminism and desire that characterizes the world of invention and *will*. Intelligence can only optimize what it has been commanded to do, but design is far more basic an affair of creative will; it gives place to what did not previously exist. Should not design today learn again to command the intelligence, not merely make public parade of its subservience to more prosaic forces?

1 Stan Allen's is by far the best of the lot, followed by Bart Lootsma, both in *El Croquis* no. 86. Others include Winy Maas, *FARMAX* (Rotterdam: 010 Publishers, 1998); Michael Speaks, "Big Soft Orange," catalog, 1998; and *ARCH+*, "Die Moderne der Moderne: Entwurfe zur Zweiten Moderne," no. 143, October 1998.

2 The Amsterdam-based UN Studio (formerly Van Berkel and Bos) is a notable exception that proves the rule.

3 "Big Soft Orange" is a traveling exhibition, not of work, but of a catalog, or rather, of a branded graphic interface, created for work that does not yet exist, but could be made to exist if marketing conditions look favorable.

Manhattan Office Building, Andreas Feininger, 1962.

THE "AVANT-GARDE" IN AMERICA

(or The Fallacy of Misplaced Concreteness)

"The Puritan wanted to live a vocational life—we are forced to."
Max Weber (1921)

The early years of the 1920s gave birth to a plethora of intense new terms and phrases: *montage*, the arraying and release of motive images in controlled, modulated succession (1929); the curiously arcane *timbre melodies* and *atonalism* of musical composition (1922–4); *Joycean*, a term at once of endearment and derision in literary circles; *quantum mechanics*, which posed as systematic a challenge to epistemology as it did to physical theory (1922); and among these others, probably the first use in English of the adjective *avant-garde* (adjective, 1925; noun, 1915). These words emerged together in mood, place and time, and they belong well together, so long, I would argue, as the words "architecture" and "America" are kept far away.

The 1920s were a time of profound changes in lifeworld and worldview, and a time of many changes that sought to appear so, but were not. Today, however, we are continually censured when we wish to speak earnestly of an avant-garde—by sterile academics able to feel intelligent only when disqualifying what they hold to be illegitimate passions possessed by others—yet to speak of an avant-garde is both a right and a political obligation. This obligation must carry us beyond the *claque* of revisionist critics and historians to principles and to fundamental questions. To speak about the avant-garde is therefore more difficult today than ever, for it now entails ethical and metaphysical dimensions, dimensions of intellectual commitment that go beyond mere sociology and aesthetics. After all, the very term "avant-garde" represents not only the idea of directed novelty (not any-novelty-whatsoever), but specifically novelty in the service of hope. The notion of the avant-garde is useful for us today precisely for its capacity conceptually to oppose these two types of novelty, these two types of historical change: on the one hand, the breaks of fully geological magnitude out of which systematic human transformations become genuinely possible, and on the other, those routinized forms of innovation that do little more than mark the onward march of technological, economic and scientific rationality.

The words introduced at the outset are each drawn from a domain of cultural and scientific research, and each marks what might be called an internally troubling development for the contexts or disciplines from which it emerged.

These diverse, individual perplexities can be said to have in common the projection into the world of a logic and a set of material givens that are irreducible to, and which fundamentally supersede, the classical framework of their respective traditions, which would normally be relied upon to give them sense. James Joyce, for example, quite simply did not write panoramic novels, and the world, according to quantum mechanics, is made not of solid particles but of fluxes and waves. At stake in each case—and I include here the projection of motion into the culture of the picture (cinema) and of atonal or dodecaphonic composition into the aural chromatic scale (atonalism)—is a radical transformation of what is seen to constitute fundamental relations in the physical world. The problem of physical existence, expressed simply in the question "How does a physically existent thing present itself to us, how is it organized, and how is it to be conceived?," goes back to the time of the Greeks (notably to the pre-Socratics, and to Pythagoras), but the question to which our own lived world forms the fully expressed answer is one that intimately concerns the status of our world's invisible but pervasive armature—embedded mathematics. The relationship between the mathematical and the physical must form the framework against which any modern understanding—aesthetic as well as scientific—must invariably be placed. For the most part, however, this is neither obvious nor is it conventionally recognized by analytical practice. On the contrary, the real strangeness of this intimate relationship is by now mostly hidden from us, and somewhat insidiously so. Arguably we would scarcely be aware of it at all were it not for the spontaneous emergence of the historical avant-gardes over seven decades ago!

The story, in abbreviated form, goes something like this. In the early seventeenth century a genuine revolution in thought took place that has come to be known as the Scientific Revolution. The Scientific Revolution brought with it a new conception of nature, one already clearly visible in Nicholas of Cusa's fifteenth-century vision of the universe as a mathematical construction of circles, centers, infinities, and spheres, and one increasingly systematized in the work of Galileo, Descartes, Leibniz, Newton, and a host of others for whom Nature would ever after appear as the ordered, mechanical expression of mathematical relationships and laws, and who themselves produced a never-ending, ever-refining stream of numerical formulas to mirror it. Nature was seen, according to the Platonic metaphysics that had newly and effectively superseded the more dynamic and qualitative Aristotelian one that preceded it, as an intricate but unmoving and changeless structure (Descartes's *res extensa*).[1] Thus as science and philosophy proceeded on their converging and world-building courses, the physical came more and more fully to be identified with the mathematical. It is this fundamental relationship perhaps more than any other that found itself radically destabilized by the cultural and intellectual developments of the 1920s. Among philosophers, Nietzsche, Bergson, and Ernst Mach had anticipated the crisis in scientific knowledge that emerged out of the nineteenth century, but only the British philosopher Alfred North Whitehead later recognized that the challenge of these transformations could be met only with a total rejection of the metaphysical system upon which modern science had been based since its formation. For Whitehead, what had emerged was a pressing, even epochal need for a new metaphysics of nature.

Whitehead's work between 1919 and 1933—from *Enquiry Concerning the Principles of Natural Knowledge* through *Adventures of Ideas* and comprising the total cosmology of *Process and Reality* and the abbreviated Lowell Lectures given at Harvard in 1925, *Science in the Modern World*—dealt primarily with the metaphysics of physical reality, the status of physical entities, and the specific problem that concerns us here, the question of what constitutes "concreteness." Whitehead's enquiry into concreteness begins with a general critique of the seventeenth century's abstractive notion of space as expressed in what he calls the doctrine of "simple location:"

> To say that a bit of matter has *simple location* means that, in expressing its spatio-temporal relations, it is adequate to state that it is where it is, in a definite finite region of space, and throughout a definite, finite duration of time, apart from any essential reference of the relations of that bit of matter to other regions of space and to other durations of time.[2]

A few years earlier Whitehead had argued that it is a fundamental —and generally unexamined—given of the schema of traditional scientific explanation in terms of Time (homogeneous), Space (inert), and Material (matter, ether, electricity), to be based on the idea of a pure, or material, extension, which has as its subtext the principle of *disconnection*.[3]

In other words, according to the modern "materialist" view, the physical world is made up of discontinuous substance, isolated points, and univocal relations. For Whitehead, co-author of the *Principia Mathematica* and a mathematical physicist himself, mathematics represented an immensely powerful tool of abstraction, allowing us to relate things of concrete fact (*relata*) that are not otherwise relatable through mere intuition or sense data. The man, he points out, who noticed "the analogy between a group of seven fishes and a group of seven days made a notable advance in the history of thought."[4] But at the same time, he stresses, in the concrete reality of immediate experience, that is, among the primary elements of nature, "there is no element whatever which possesses this character of simple location."[5] It is not, of course, that scientific expression is in error; it is simply that science in its essence is a technique of "constructive abstraction." We can indeed, and without inordinate difficulty, arrive at simply located bits of material, *but only in and through constructive logical abstraction*. The problem is that we can never arrive at such simple location in the naturally occuring physical world itself. The fallacy to which this situation almost invariably gives rise, he argues, and the one into which the seventeenth century and the subsequent tradition of scientific enquiry categorically fell, was, *in their metaphysics*, to have mistaken what is abstract for what is concrete.

To this conundrum Whitehead gave the label: The Fallacy of Misplaced Concreteness. In essence it decries the systematic self-delusion on the part of a lengthy, seminal philosophical tradition, the fallacy that space and time are simple, inert, and serve as a detachable substrate to the events which take place on or within them. The theory of motion, for example, was not an easy problem for the seventeenth century. At the very least it necessitated a coherent and defensible theory of *place*, a datum as it were, in relation to which the concept of movement could be extracted as an abstractable numerical value. Thus the mechanist Descartes could do no better than to describe motion as "the transference of one part of matter or body from the vicinity of those bodies that are in immediate contact with it, and which we regard as in repose, into the vicinity of others."[6] What for Descartes was still just a vestigial potientiality in passive, indeed inert, "repose"—he had, after all, not fully abstracted the difficult concept of place from things—would later, in the hands of Newton, become a total conceptual and immaterial construct, that is, an *absolute*, and a divine one at that.[7]

Motion was stripped of any of the qualitative, transformative values with which it was still endowed in the Aristotelian-medieval worldview; its effects were now sequestered within the numerical construct that descriptively linked discrete figure to impassive ground. The schema of the seventeenth century was in this way completed: the enormously efficient feat of abstraction had yielded a functional polarity: *Matter* with its simple location in absolute space and time on one side, and *Mind* (*res cogitans*) on the other, active and functional but entirely placeless and distinct. This, according to Whitehead, is what our philosophy and popular metaphysics have inherited: a false dichotomy, sterile dualisms, and the endless confusion introduced by the ascription of misplaced concreteness to what is only the modern scientific epistemological scheme. The Fallacy of Misplaced Concreteness, and the culture of inquiry that grew around it, has laid any possible apprehension of real physical reality beyond the scope of modern thought. Between the terms of this disfiguring dualist (matter/mind) abstraction, however, Whitehead writes, "There lie the concepts of life, organism, function, instantaneous reality, interaction, order of nature, which collectively form the Achilles' heel of the whole [modern] system."[8]

What then is the actual physical existent? Where does real concreteness lie? It is not hard to guess from the assertion above that it may be found neither in extension nor in mind, neither in space nor in time—for these are but technical abstractions from more primary dynamic phenomomena:

> The actual world is a manifold of prehensions; and a "prehension" is a "prehensive occasion;" and a prehensive occasion is the most concrete finite entity, conceived as what it is in itself and for itself, and not as from its aspect in the essence of another such occasion. Prehensive unification might be said to have simple location in its volume *A*. But this would be a mere tautology. For space and time are simply abstractions from the totality of prehensive unifications *as mutually patterned in each other*. Thus a prehension has simple location at the volume *A* in the same way as that in which a man's face fits on to the smile which spreads over it.[9]

What does this tell us? First, that "concrescence" comprises the primary activity and fact of nature, that nature's concreteness may legitimately be found nowhere else than in its *activity*. Nature is first and foremost *process*,

No contemporary practice has made so foundational an argument about the skyscraper and contemporary space as has Abalos & Herreros—not OMA, not MVRDV, not even Le Corbusier before them—and in this the genius of their work lies. The skyscraper is presented first as a system of knowledge (*Tower and Office*, MIT, 2005), then hypostatized not as a building type but as a piece of public and social equipment. While others proclaim the age of the skyscraper to be over, Abalos & Herreros respond that *it has not even begun* . . . Indeed the skyscraper, they show, is our fate. Today the skyscraper represents no longer principally a visible figure—as in the mid-century American heyday—but an environment, indeed a type of *utility*, to use a term common in the contemporary business environment. Utilities denote infrastructural entities like water and electricity that are perceived to be always there and upon which we depend without thought or specific awareness or action. Like a Trojan horse, the skyscraper can be said to have smuggled its own set of functional ubiquities into the contemporary social and economic world. No longer an autonomous centripetal "condenser" as in the immediate postwar City, it is today rather a relay node within a much broader ("global") system of social dispersion. The late, "neo-" modern, or "second wave" skyscraper is arguably the social counterpart to the advent of numericalization and automatic processes in our technical world, providing for the second time within a single century the purest image we have of what Max Weber called "a precision instrument of formal, technical and political rationality."

The classical age of the skyscraper was the age of the business Corporation. Today's skyscraper belongs to the regime of machines and environments made possible by the marriage of computing and communications: it is spatially and technically generic and all friction has been removed from its internal organization (a veritable *Infolandschaft*). The removal of obstacles means both the removal of remaining structure and its exportation to the building's exterior (envelope), but also the incorporation of the new logics of behavior and use that give pragmatic effectiveness to these new "spaces without qualities." In other words, this evacuation is made possible because these new buildings have been saturated with ubiquity, a ubiquity of services (from air management to forests of data feeds and even natural parks) that makes possible an almost infinite adaptability and flexibility—in other words, cryptic post-Fordist locales. Abalos & Herreros argue that the contemporary skyscraper, whether it is literally present or not, is not simply a mirror of our city, but *is* the city itself.

For this reason it is safe to say that every building of Abalos and Herreros develops a skyscraper. Every work consciously and deliberately "redescribes" a form type (redescription is a central term for them, derived from American philosopher Richard Rorty), the form type that characterizes what they call the "contemporary topological condition." Even the diminutive two-story Retiro Park gym depicted here rhetorically projects a curtain wall of uniform modules upward *as if* . . . to infinity. The space within is invested with the typical mixture of heterogeneous uses—gym, dressing room, offices, plus an elevated roof tennis court and most importantly a multitude of trees intended to adhere to, and over time to digest and obliterate, the mesh curtain wall surface around it. It is in this last gesture that we discover the core tension that saturates their work: that between the "Faustian" aspect of human life, the wild, free and spontaneous aspects of nature and *individual* existence—in a word, the irrational—and the historical machinery of rationality that has been put into place ostensibly in order to free us in our *collective* lives. Within the luminous rectilinearity of Abalos & Herreros's work there is always centrally posited a certain relationship to nature and freeformality, often violent, never vanquished, and never suppressed. Every work is "an urbanism" that plays out a Sisyphean drama in which Modernism's drive to emancipation is relentlessly undone by the limits of its own implacable rationality.

or rather, it is a manifold of evolving processes, here described as the ceaseless formation of prehensive unifications. By definition prehension consists in the very act of forming actual unities from all of the possible perspectives distributed within the world's prehensive manifold. Unification should be understood here in terms of "connectedness," the type of connectedness that may be said to hold between the elements that come together to form a single event.[10] Thus the event in Whitehead's cosmology may be understood as "the emergence into actuality of something." The concrete "event" in Whitehead's cosmology is meant to replace the abstract notions of space, time and object, while the concept of "organism" is made to represent the "concrete enduring entities" that are the full expressions of all concrete occurrence. Whitehead shows that the classical space and time of the modern scientific era are at bottom techniques of separation: things are separated in space as they are separated in time. But things are also *together* in space and together in time, and it is to this aspect that the "prehensive character" of space-time refers. In addition to the "separative" and "prehensive" characters, there exists yet a third, the "modal" character. The modal character concerns what I call a substrate activity of nature or, to speak more loosely (and germanely), of *space*. This substrate activity has to do with the impossibility of speaking strictly about a point in space-time (simple location) without reference to its embeddedness, that is, its *relations* with other points both distant and near. After all, as Whitehead points out, a volume cannot be conceived of simply as made up of a multiplicity of points, because points are by definition empty, and a volume is manifestly a concrete fact of experience. Of what, then, is it made up? The answer is that a volume is made up of parts, though not in the sense of a simple aggregate to which things are merely added, but as an entity compounded of a selection of many aspects that impress themselves (within the prehensive event) into every other part. Yet the true concreteness of the volume is not here either, but in its evental appearance as organized or "patterned" at a given place: If *A* and *B* are parts of the volume *C*—for example, the room in which you currently find yourself—*A* may be understood eventually as presenting an aspect of itself to *C* just as *B* does. But a specific relation also exists between *A* and *B*, again made up of that particular aspect that each volume may be said to have from the standpoint of the other. This relation *AB* must also be understood as impressing its contour into our volume *C*.

The aspects of A, B, and AB from the standpoint of C contribute to, or partake of, the essence of C by means of what is called their *modal ingression*. Not every possible aspect of volume A or region B is actually active in C, but the degree to which some of their potential aspects do ingress, or to which they *enter into composition* with volume C, they may be said to be *individualizing* entities rather than simply individuated ones. The important thing to understand here is that they are actively expanding and becoming when grasped or prehended into unity, which is to say, when they burst into actuality by taking on and contributing specific (selected) determinations. The actual(-ized) unity is therefore an *acting* or an occurrence, it is the organism-event understood as the concrete structure or scaffold of interlocked evolving processes. To extend Whitehead's example of the smiling face, what we have here is the reality of the smile onto which the face is caught, not that of the face from which, in the habitual and disfiguring (so to speak) materialist mode, it would otherwise be said to be abstracted.

In the final analysis, concreteness consists in the process by which general "simples" become realized or expressed in specific prehensive events. The connection between regions of the physical world, of nature, and between the substrate of so-called "eternal objects" that lie awaiting determination through actualization and the prehensive occasions which receive them, is *real*. This is the second, and by far most difficult, meaning of the Fallacy of Misplaced Concreteness. According to this doctrine, metaphysics has mistakenly ascribed concreteness to compound entities rather than to the substrate *activity*, that is, to the process of realization or to the "becoming" which shaped them. But the unity of the event is more directly derived from the multiple ingressions or the plurality of modes. More simply, concreteness is rooted in internal relations.[11]

Every prehensive event is an organism. As Whitehead put it: "Biology is the study of the larger organisms; whereas physics is the study of the smaller organisms." What he means is that the plan of every whole influences the character of its parts by means of the given pattern, the *routes*, by and through which they interact. The parts in an organism are continually *acting upon* one another. Their relations with one another actually contribute to, and determine, their respective essences. The unity of the prehensive event is a seizing together of aspects or relations between things;

Prehensive occasion? Korean Presbyterian Church, Queens, New York, Greg Lynn FORM, Garofalo Architects, Michael McInturf Architects, 1999.

everything is determined—by deformation and limitation or selection—by everything else. This interfusion and mutual emendation of events through their compound relatedness is in fact our elusive real concreteness in its most simple form. Metaphysics has failed to ascribe concreteness to this, the most fundamental mode of being of the physical world.

It is the failure to recognize the primacy of relations, and the failure to recognize the substrate process or agency that ensues from these, that constitutes the Fallacy of Misplaced Concreteness or the failure of traditional metaphysics to account for real physical existence in its dynamic heterogeneity, organicity and multiplicity. The very concept of an order in nature could, for centuries, not even coherently be posed, let alone the theory that nature might first and foremost be conceived as the locus of organisms in the process of development.[12]

***

By the 1920s a new movement, mood, and style, and many lucid proclamations, had emerged in architectural milieus in Russia, France, Italy, Germany, Holland, even in America. The European movements often, yet not always, had roots in political and social developments, but none failed to engage revolutionary ideas of cosmology derived from physics and philosophy, or the social sciences. Architecture sought emancipation from restrictive historical programs of all types, while modernization was embraced as if it were a salvational religion. Technical innovations, while not the cause, certainly provided much of the fuel—material and intellectual—that drove the movement, and helped confuse what the word "modern" would thereafter come to mean. The Modern Movement had many chapters and many parts, but collectively, at least for a time, it may be said to have formed a coherent project: the search for new fundamental principles through which the physical world might appear both to us, and for us, as open, changeable, and free. But this program and its clear aspirations cannot be separated from adjacent social and civil developments on the broadest possible scale: to wit, the systematic rise of the modern scientific bureaucracy, the appearance of what the great sociologist of religions and economic historian Max Weber would call that "precision instrument" of formal, technical and political rationality. Though there is neither space nor need to present a full analysis here, I will take up two points that relate to Whitehead's problematic, and to the fate of modernity in America.

Modern bureaucratic organization as it arose in the nineteenth century, and as we have come to understand it through the analyses of Marx as well as Weber, may be said to be based on the two principles of simple location, that is, on formalisms of separation and abstraction. Bureaucratic organization entails separation in the sense that it both effectuates and exploits what Marx called the separation of "producers from the means of production," only here a more complex and nuanced *civil* separation is implied—that of civil servants and specialized administrators, from the means of public administration. Abstraction is played out more obviously as the abstraction of concrete, spontaneous—what Weber would occasionally call "Faustian"—humanity, first into degraded *Berufsmenschen* (specialized or vocational men, or, etiologically, men subject to divine obligation or a "calling"), finally reducing them through abstract regulation and rational calculation to the status of mere administrated entities or "files." The spread of bureaucratization is inseparable from modern capitalism (indeed, its roots are commonly associated with the rise of the money economies from as early as Roman times), inseparable from the secular rationalization process which in the West has focused powerfully on the domains of politics, industry and most notably, the law. Yet these processes, which can be reduced to a single idea—the rationalization of action—have penetrated in an almost unfettered manner into all spheres of culture, including music, literature, and, over and above all others, architecture.

Now, "the rationalization of action" is not an idea to be passed over quickly. It tells us that there exist formal entities in our cultural world that are capable of inducing strictly precise behavioral routines and codes of conduct. We are not surprised when told that religious doctrine or systems of belief belong to this category of institution, but we are more surprised to learn that formal structures of social organization as mundane as public administrative apparatuses draw upon, and organize, these same "irrational" or magical forces.[13] Modern bureaucracy, which Weber himself compared to the advent of the industrial machine in recognition of its clear technical superiority over all previous systems of production/organization, is primarily oriented toward submitting social conduct to rationally calculable principles. Such an operation does not merely assume that both objects and humans behave in equally predictable, quantifiable ways, it firmly cultivates and induces such types of predictable behavior. It is, of course, only the *ethic* of Protestantism that is required in order to forge a *spirit* of capitalism; in other words, it is a relationship of dynamic formalisms (even if life-hating ones) whose substantial prescriptions and constraints remain for the most part atmospheric and invisible.

The "rationality" of bureaucracy, among other things, lies in its arguably democratic capacity to regulate *generally*, to transform every *individual* case into an abstract member of a *category*. The reduction of specificity, and the meticulous application of a fixed, preexisting regimen of administrative protocols ("the exact execution of the received order"—Weber) is the social equivalent of numericalization (quantification) in industrial and experimental scientific milieus. Indeed, bureaucratization, insofar as it is rational, is the crowning result of the wholesale penetration of modern science and its mathematical substrate into the social sphere. The transformation of magical or religious belief into the decidedly disenchanted but highly effective rationalistic posture toward nature and world that characterizes modern capitalist organization represents, however, only a redirection of faith, not its destruction.

Modern social life—with its telephones, high rises, therapists, ergonomic imperatives, actuarial tables, marketing and entertainment apparatuses, etc.—is fraught to the point of saturation with scientific principles, yet their equations need not be explicit or visible to us for our conduct in the world to be entirely governed by our implicit faith in their existence. The world- and event-continuum is quite simply no longer even remotely believed to be made up of mysterious and intractable anomalies, special cases, and magical forces; on the contrary, everything—even the flying circus of human life—is seen to be calculable and hence masterable by virtue of that very calculability.

It is not by accident that the theory of the division of labor is routinely used to describe both modern bureaucratic technique and the advent of the rational juridico-administrative apparatus. All are based on the actual *instrumentalization* of the Fallacy of Misplaced Concreteness and its system of accompanying reductions. Weber continually decried this process as the virtually unresistable tyranny of the enforced "calling" (that Protestant distortion in which we moderns are compelled to recognize our lives) and the ugly, irreversible ascendancy of "the bureaucratic ideal."

***

In the 1930s three Americans with position and means set out for Europe in search of a style.[14] Like proverbial rubes in the big city, however, they quickly fell dupe to the cardsharps of history, and unwittingly acted out the very script that Alexis de Tocqueville had written for them a century earlier. What they brought home was neither European nor even remotely "advanced," but rather a hodgepodge of enthusiastic but mostly trivial insights and frivolous infatuations that amounted to no more than reorganizing and aestheticizing a broad set of plastico-material practices whose specificity, meaning, and historical purpose escaped them, yet whose clear, uninflected forms were easily amenable to the cultural logic of the simple, the efficient, and the slightly sadistic but taut mathematical beauty of administrative space. They brought back to America, in other words, pre-packaged forms that seemed a perfect fit for what American society had already programmatically forged: "elegant" (one of their favorite terms of approbation and ironclad legitimation) correlatives of the numerical space of Taylorist "scientific" management and of the deadly efficient, homogenizing space of Fordist mass production/mass consumption loops.[15] The "international" modernism that they brought home was unconsciously tendentious and may be seen, in the absence of any alternative critical apparatus of reception at the time, as a vulgar aesthetic alibi for a nascent yet burgeoning new type of cultural institution: the industrial-administrative corporation. The business corporation is a peculiarly modern entity, not merely a huge *Fabrik* or manufacturing enterprise, but a massively specialized society of administrative, managerial, and productive technologies that includes both implicit and explicit systematized rules of personal conduct, new architectural forms such as the huge, continuous upright curtain-walled fields of undifferentiated offices and files, and a program of uniform global self-presentation within the public sphere. This latter development is itself perhaps the most sophisticated and scientific of all bureaucratic activities: the communication and engineering of what has come to be known as identity.

While Europe wrestled with these same developments (more highly evolved corporate entities already existed in Germany and Switzerland), its aesthetic and architectural movements, their representatives and their thinkers, never detached themselves from the dynamics of crisis: social, political or epistemological. It is known that the American shopping expedition rejected works that bore the taint of "expressionism" even when steeped in the most advanced speculations regarding physical matter in movement, tension or decay: Mendelsohn's extraordinary "spasm" of 1919–22—the *Einsteinturm* in Potsdam—is one egregious example, as are the startlingly inventive but more heterogeneous experiments of the Soviet Constructivists. It is no surprise that Lewis Mumford, the one deeply civilized and cosmopolitan imagination to have been involved in the planning of the International Style exhibition, was compelled to withdraw from the project on ethical principles. Americans, as de Tocqueville well knew, were simply not (yet) interested in ideas.

In assimilating the progressive modernist research that was inseparable from Europe's struggle to emancipate itself from traditional forms of social domination, to the pragmatic but predatory logic of the emerging international corporate agenda, that is, to the bureaucratic ideal, an entirely retrograde space was put at the service of social engineers, and the speculative imagination of America's young and spectacularly inventive culture was stifled and shunted back to the sterile utilitarian mechanism of the seventeenth century. In other aspects of its culture, in its literature, music, biology, and for a few decades, in its cinema, new languages and material intensities began to proliferate steadily, while the New World's architecture remained merely elegant, effete, brutally corrupted, and, if not entirely retrograde, then certainly naive.[16] Architecture in America, because of the untimely meddling of dilettante esthetes, served in subsequent decades as a Trojan horse that in both visible and invisible ways hobbled the American social and aesthetic imagination, even while its cities and its industry were becoming the marvels of the western world. The false and bland concreteness that these vain yet grand technological monuments passed off as harmonious and universal pattern was little more than what Foucault would have called the *diagram* of the invisible, routinizing cage within which modern humans unconsciously yet unhappily are forced to live their increasingly petty, bullied and attenuated lives. Clearly I refer here not directly to the physical works of architecture themselves, nor to the alternative possibilities that undeniably lay poised within them, but to the physical argument or modulus that they actually transmitted into a complex historical calculus of motives, rationales and instrumentalities. The question is not whether there was a systematic avant-garde spirit in America in the middle decades of the century, but whether the American spatial and physical imagination can yet possibly claim to have recovered from the confusion and ineptitude that had earlier been visited upon it.

1 The concept of physical existence in Aristotelian metaphysics was based on actualization—energeia—of inbuilt potential, and on *kinesis* where motion is conceived as a principle of internally driven change, transformation or becoming residing within the existent itself, and which is not reducible to mere translation or change of position.
2 Alfred North Whitehead, *Science in the Modern World* (New York: The Free Press, 1925), p. 58.
3 Alfred North Whitehead, *Enquiry Concerning the Principles of Natural Knowledge*, chapter 1, section 1.2.
4 *Science in the Modern World*, p. 20.
5 Ibid., p. 58.
6 *Descartes, Principles of Philosophy*, part 2, principle 25.
7 Newton could not bear the idea that the vicinities or places themselves might be movable because this would render precise measurement impossible. In the first Scholium of his *Principia Mathematica* he grants the existence of absolute, immovable places and calls the sum of these places, space (*spatium*). Yet ascribing real "concreteness" to this mathematical construct was, curiously, for him philosophically untenable. He is thus led to summon a God to serve as the creator of all places. The activity of his ceaseless creating becomes the universal substrate he calls "space." See Ivor Leclerc's strikingly original analysis in *The Philosophy of Nature*, vol. 14, *Studies in the History of Philosophy* (Washington, D.C.: Catholic University of America Press, 1986), p. 140.
8 *Science in the Modern World*, p 57. Whitehead claimed that his philosophy was an attempt to produce an "organic mechanism."
9 Ibid., p. 71.
10 *Enquiry Concerning the Principles of Natural Knowledge*, chapter 1, section 1.5.
11 The "internal relations" school of philosophy is associated with the work of F. H. Bradley, J. McTaggart, and other British neo-Hegelians of their generation. It seems clear to me that Whitehead's work issues from theirs, though I have seen no confirmation of this at the present writing.
12 *Science in the Modern World*, p. 73.
13 This is Weber's most famous thesis. See *The Protestant Ethic and the Spirit of Capitalism* (New York: Scribner and Sons, 1958).
14 Alfred Barr, Henry-Russell Hitchcock, and Philip Johnson. Possessed undeniably of genuine curiosity and zeal, they were apparently the first Americans to feel at least professionally concerned by new European developments. They were all, however (with the partial exception of Barr, who possessed

significant understanding of and not only acquaintance with European painting), too deeply entrenched in the nineteenth century parlor traditions of connoisseurship and refined appreciation to sustain any concern with social historical or philosophical aesthetic issues. Hitchcock's writings, for example, particularly his defenses of the International Style exhibition that was the ultimate issue of their trip, are comical in their confident displays and defense of vapid sensibility over principle and idea, all the more so when read against the intellectual standards of modern times. We now know, from Franz Schulze's biography of him, that for a time Philip Johnson became an acquaintance of Whitehead and his wife during his undergraduate years at Harvard despite failing to impress Whitehead intellectually or even to pass his course successfully. It seems that this was an important turning point in his life, since it convinced Johnson that he was not particularly well endowed for intellectual work and so could more usefully turn his energies and attention to promotion, alternately of fascism and design. Johnson, who later signed and paid for the exhibition's essay, and to which he apparently contributed some editing, was, by his own admission, primarily a booster along for the ride.

15 The Model T of 1914 represents a quantum leap in the bureaucratic mastery of social routines: workers are transformed from producers to consumers, the market penetrates even the factory floor, and workers' wages are recycled as profit through the repurchase of products they themselves produce. The increasingly motorized workforce permits the steady transfer and refinement of mechanical expertise from workplace to leisure modalities and back in a mutually reinforcing loop that will soon become so encompassing and constitutive of our world that it will disappear from view. Optimization of productive locales (factories) in relation to housing, secondary and correlative markets, and the total reshaping of urban environments according to rational economic principles overcame a major inertial obstacle in social space by means of the so-called democratization of mobility with the af*ford*-able (as it were) automobile.

16 Rem Koolhaas's appealing, career-long apologia for this architecture notwithstanding, it represents less a recovery of, than a corrective to, the tradition legitimized by the Museum of Modern Art exhibition group. Koolhaas is a European's salvaging of a European tradition that Europe, in its weariness, could not sustain alone. America's modernism, one might say, was Europe's *Aufhebung* in action.

IBM

Movable type, the sequel: The Selectric Universe. IBM Pavilion, 1964–65 World's Fair, New York, Charles Eames and Eero Saarinen. Culture's last image of alphabetic causality before the advent of the graphical user interface.

The "mechanical" and the "electronic," contrary to popular cultural theory and belief, are by themselves *not* paradigms and do not represent distinct, successive, agonistic "ages" or irreducible worlds in collision. To continue to think of these notions in such worn and sterile ways can have no other effect than to hide from ourselves their more important and nuanced political dimensions. The mechanical and the electronic, and most of what is anxiously denoted by these terms in present usage, are in fact expressions of two continuous, interdependent historical-ontological *modalities*: those of Matter (substance) and Intelligence (order, shape). Every unit of intelligible matter in our technical or cultural world, regardless how simple, is refined or organized to a degree sufficient at least to distinguish it from the random and disordered background flux or noise of the natural world. (Of course, natural objects may possess this same property of refinement in proportion to how closely they are formed and organized by the processes of *life*, processes now commonly thought to extend beyond the merely organic.) In this sense such matter may be said to possess a greater or lesser amount of embedded intelligence. By intelligence one can understand a set of instructions accumulated over the ages (through the application of tools and controlled processes) and now incorporated into the matter in question as a permanent and continually reactivated "memory," that is, either through shape, rhythm or disposition, as in a tool, or through purity or precise proportion, as in the relationships of metals in an alloy and the properties that derive therefrom. All matter, even totally disorganized matter, possesses some degree of active intelligence (what Diderot called "sensitivity"), and the refinement of matter is always the refinement of the intelligence (shape or the proportions and patterns of its composition) embedded within it. When different types of matter, or different orientations of intelligence in matter, are brought together in a proper or sympathetic rhythm or proportion, an entirely new level of intelligence is given place. As the complexity of added and engaged elements increases, one approaches, then arrives at, the mechanical.

The mechanical is thought to be primitive in perhaps two senses. First, its relationships of intelligence are based on rudimentary, visible associations of isolated elements that interact at a very reduced level of information exchange. What this means is that in mechanical devices,

THE CRUELTY OF NUMBERS

most qualitative information tends to be reduced or eliminated in favor of a very controlled, exclusive extraction of quantitative flows (rates of movement, measures, etc.). Most elements in a mechanical complex have a single function and a single, circumscribed set of relationships. In other words, the very purpose of their design is to suppress most of the material-embedded intelligence in order to favor or isolate a single quality or dimension of expression. It is here that the commonly confused terms *mechanical* and *mechanistic* find at least a momentary and justifiable convergence. (The first term means having to do with machines, the second means deterministic, and reductionist.) Second, the mechanical is thought to be primitive in relation to electronic processes, because the latter appear to manifest the same magical qualities of *material intelligence* found in fundamental, free and unprocessed matter, a set of qualities that can be summed up in the term, *self-control*. Control here means simply the sustained application of intelligent—or organizing—force over time. While the mechanical complex seems over-designed and therefore limited to very rigid and predetermined pathways of unfolding (i.e., no development at all), archaic and electronic matter-complexes are thought to be able to move and evolve in coherent yet unpredetermined ways. Their manifest intelligence is both multispectral and freeform, or "complex."

All (advanced) technological societies have evolved from archaic matter intelligence (empirical, qualitative, multispectral) to mechanical matter intelligence (numerical, dissociated, linear), but only incompletely and each in its own way. In the West, mechanical matter intelligence took on an almost religious status (as electronics is certainly achieving today) to the point of annihilating archaic matter intelligence from public and social memory. The way in which each society organizes its systems of intuition—its science, its philosophy, and its technics—is in every manner a political one. The real and possible arrangements of intelligence and matter in nature are one thing (and likely unlimited), but how we represent these possibilities to ourselves is another. To speak of a mechanical paradigm of material qualities and perceptible functions, and to oppose this to an electronic one of immaterial processes and pure intelligence, is at once absurd and dangerous. Absurd, because despite what cyberspace gladhanders pretend to think, there can clearly be no shape or order (Intelligence) without matter,

First published in *ANY* 10, "Mech-In-Tecture: Reconsidering the Mechanical in the Electronic Era" (1995).

Patek Philippe Calibre 89, the world's most complicated watch. This watch, the king of the "complication" category of tour de force watchmaking, contains thirty-three mechanical complications, including one that calculates the hour of high tide and a display that calculates the changing date of Easter until 2017.

even if this matter is composed of nothing more than pure photons (cinema, retinal laser projection) or molecular acoustic resonance (music), and dangerous, because such oppositional clichés do little more than render us stupid and docile in the face of disfiguring yet well-camouflaged social and historical processes.

What is taking place today is nothing to do with the eclipse of a material or mechanical world by an increasingly electronic one, but rather the emergence of a new regime of subjection that uses the undeniable allure of an archaic revival (a return to matter, complexity, and free development) to facilitate a repressive reorganization of social space, as well as a mastery of the very conceptual lexicon with which this reorganization will be thought through. More bluntly: what is occurring under the guise of "rational" historical process is the systematic formation of a new subjectivity—a new type of "Man," to use Nietzsche's expression—whose matter/intelligence variables are being reengineered and finely calibrated to fit those of a new machinic workplace-society into which s/he is to be seamlessly integrated.

It is therefore disconcerting to observe the direction of much discussion in the design world around the advent of new telecommunications technologies, computerization, and software-driven milieus. These developments are either extolled as "exciting," "new," and "full of new freedoms and possibilities" (by those blissfully unconcerned that much of what is being so celebrated is but an extension of all that is oldest and most repressive in our political and corporeal history), or else are seen as posing an unavoidable or even welcome challenge to an already weakened or near-obsolete domain of cultural practice, namely the slow, grave, viscous world of matter. The routine disdain heaped on matter by both these points of view is in fact focused ideologically on an officially distorted notion of the mechanical—made now to mean anything that is concrete and available to intuition. Our task, I would argue, is to resist these pathways of thought, and wherever possible *to expand the concept of the concrete* and *to extend the play of intuition into new domains*. To do this effectively, it must remain within our power (conceptual and political) actually to refuse the advent not so much of the specific machines and techniques of contemporary development but of the broader systems of rationality in which they come packaged or for which they serve as Trojan horses. Communications networks, computers,

Cogwheel milling machine from 1794 used to create proper watch teeth with precise ogival form, from Patek Philippe Museum, Geneva, Switzerland. Antonio Gramsci argued that a society's technical evolution could be measured by the sophistication of the machines that it creates to produce other machines.

microprocessor control systems are socially toxic entities primarily when used "correctly," that is, in their capacity to routinize interactions with people and processes in increasingly engineered, confined, and deterministic spaces. It is our duty and mandate to refuse this new, pseudo-material space entirely, and to follow the "minor," *archaic* path through the microchip, that is, to make the electronic world work *for us* to reimpart the rich indeterminacy and magic of matter out of the arid, cruel, and numericalized world of the reductionist-mechanical and the disciplinary-electronic.

No computer on earth can match the processing power of even the simplest natural system, be it of water molecules on a warm rock, a rudimentary enzyme system, or the movement of leaves in the wind. The most powerful and challenging use of the computer (aside from the obvious benefits of automated number crunching in purely numerical domains such as astronomy and accounting) is in learning how to make a simple organization (the computer) model what is intrinsic about a more complex, infinitely entailed organization (the natural or real system). Implicit here is the idea of learning how *to make matter model matter*, or how to study natural or "wild" intelligence in a contained but active, refining domain. In this use the computer becomes *metallurgical* substance, it extends the exploratory, evolutionary process of differentiation and refinement by inventing new levels of order and shape. The computer and its software together can form a Matter/Intelligence unit of a very primitive but useful kind. But to do this, the computer, in the triad Nature-Mind-Computer, must play only the appropriate intermediary role of interface between Nature and Mind. This would be in clear contradistinction to what is more often the case today, where computational environments provide a customary but imperceptible experiential envelope from which Nature (and all nondeterministic unfolding) is excluded, and within which the activity horizon of Mind is insidiously confined. We must not believe the narcotizing hype that an emerging electronic world is poised to liberate us from a mechanical one, nor even that there exists an electronosphere fundamentally discontinuous from the mechanosphere that has formed us till now. It is true that an important transition is taking place: mechanical relations are being dramatically transferred to new and different levels—like the little ball in a scam artist's shell game—but they are certainly not disappearing.

What is more, however, this transition state is an unstable one, and one of the possible arms on history's bifurcation diagram (the one that does not lead smoothly to the total routinization and economic subsumption of the human organism) leads at once to the possibility of multiple new ecologies of human existence and to the dark, possibly unfathomable mysteries of nature itself. What we need today is twofold: On the one hand, *resistance*—to direct our theoretical activity away from simple-minded clichés in order to conceptualize the proper materiality of the electronic and the brutal effect it is having on both human energy and the physical environment; and on the other hand, *productive affirmation*—to actively press computation toward its deep rootedness in the archaic world of natural intelligence, which means at the very least to use computation just as the early moderns used the telescope and the microscope: to engage aspects of nature whose logic and pattern had previously remained ungraspable because they were lodged at too great a remove from the modalities of human sense and intuition.

During the Renaissance, specific movements and structures of astronomical and microscopic scale were for the first time brought into the purview of human thought and perception, and in the process certain forms of historical tyranny became forever impossible. Today the computer offers the possibility of apprehending developmental patterns of extraordinary and unprecedented depth and abstraction, offering tantalizing glimpses of the very freeform structure of time itself (chaos, complexity, self-organization). Just as Lucretius's hydraulic hypothesis in his ancient *Treatise On Nature* once proposed to free humans from the capriciousness and prejudices of the gods, so this new tool—among all the horrors and deprivals to which it is already giving place—may well bear the potential to unlock the door on the universal laws that govern the appearance and destruction of form, and in so doing, free us from the multiple tyranny of determinism and from the poverty of a linear, numerical world. Yet there should be no illusions: the possibilities for such a scenario are almost already foreclosed, and it will certainly not come to pass with anything short of a colossal, sustained, and collective act of human will. It is we, the engineers of human environment and activity, who bear the burden to ensure a properly human pleading in this struggle for our fate.

COBBLE STONES BENEATH THE BEACH . . .

Andon Board. Workers hit buttons to illuminate regions on the board that signal to management when assembly line speeds are beginning to overwhelm the "human factors." Management's opposite task is to maintain the greatest saturation of the board while keeping productivity on the lines just this side of breakdown. The Andon board is to post-Fordist production what the canary was to the age of the mines: a boundary marking the limits of tolerance, once strenuously avoided, now ruthlessly embraced.

The central concern of Bernard Tschumi's work has always pivoted around the concept of "performance": For Tschumi, architecture is always a response to social "scripts," always a subversion of the economic equilibria, always the incarnation of a public *act*. Best known for his Deconstructionist parries of the 1980s—especially the epoch-making Parc La Villette in Paris—his was the only non-Protestant reading of the work of Jacques Derrida in the architectural world, a fact that allowed him, and him uniquely, to bypass the literalisms that plagued Anglo-Saxon practitioners. Tschumi worked within an idiom that saw the source of the contradictions that animated architectural work not in form alone but in the *social* realities that engendered it; his principle device is the "mise-en-scene" of historical contradiction (*never* its resolution). His buildings represent Brechtian snapshots of the state of our turbulent social world at a given historical moment. But he also has a foot firmly planted in the French engineering tradition, from which follows his persistent interest in the infinitely subtle social meanings of glass. In the work represented here, these two tendencies are forcefully combined in a simple, taut structure in which the exaggerated discipline of the curve is interrupted by the asymmetry of the broken torus: it is in the gap between the concentric trajectories of the arcs where—just like in filmic montage—*everything happens* (it is also what the Greeks would have called the work's "aporia"). In the case at hand, the carrier of this philosophical weight is the entryway, trigger and focus of all of the work's circulation. The piercing mast and its guys (tension cables) provide a startlingly smooth, even serene, reminder that technological mastery is but a rigging and a ruse that must not to be taken as a false reenchantment of the world. Deconstruction as it was meant to be: a philosophy of *function*, not simply form.

Telecommunications bundle in One Wilshire, 2002. This building in downtown Los Angeles was built as an office tower by SOM in 1966. It has since been refurbished (cables and servers now replacing offices) to serve as the Western United States' principal telecommunications hub. It is thought to be the most expensive real estate in North America.

THE WIRING AND WANING OF THE WORLD

This lecture was written just as the author (and many of his colleagues) were setting up their first email accounts. One task of the paper was to accumulate the most ominous perspectives on contemporary social development in fulfillment of what I held to be an intellectual's obligatory exercise of resistance. In the years that have ensued since its original writing every one of the developments described herein has seen a development that has exceeded what passed as hyperbole only a decade ago. All that dates the essay today is that some of the issues raised seem inexplicably quaint by contemporary standards. For this reason I have chosen to publish the piece without further updating: it stands as a piece of historical evidence not only of how quickly the technological landscape is changing, but of how quickly the intellectual and perceptual ecologies within which we conceive of freedom and life mutate and vanish not only from view, but from comprehensibility itself.

Prologue

As members of the design profession we are easily and rightfully fascinated and inspired by innovation, seduced by the aura and grace imparted by successful and daring formal propositions, and in this, in many ways, lies the real beauty and health in what we do. Yet there is a growing tendency to become overly mesmerized by these factors, a tendency to lose sight of the *existential* aspects of our relationship to the material and built environment. This loss of existential dimension can have no other effect than to impoverish our experience of the world, to obscure our natural instinct for freedom, even to diminish the pleasure we are capable of taking in form. These three things—experience of the world, instinct for freedom, and pleasure in form—are profoundly linked, yet the existential domain in which their deeply intertwined connections are forged, and what I fear to be its current waning, will be an enormously difficult thing to reconstitute. In my own work and life, attempting to approach this problem has continually entailed the process of memory.[1]

Introduction

What I will say here and what I would have liked to say must remain separate things, because what I would have liked to say, for all practical purposes, is no longer truly expressible at all—not, at least, in our *modern*, public language of ethical and philosophical debate. The problem here lies in the unavoidable use of the concept of "modern." For this familiar term, which has

First presented on March 30, 1994 at the Museum of Fine Arts in Houston, for a lecture series on "The Virtual City" sponsored by The Rice Design Alliance.

expressed the hopeful essence of our specific type of civilization for almost five centuries, does not, as common parlance suggests, mean simply "new," but more precisely, *emancipated*: emancipated from the episodic whims and prejudices of "mere" tradition and belief (those affects typically said to characterize the premodern universe), and emancipated from the indifferent, adverse and wild unfoldings of the natural world. The essence of the modern "type" of human that corresponds to this worldview is in turn expressed—to itself and in its own vision—as a creative freedom struggling in and with an indifferent world to shape itself as it wills. That human *will* and human *freedom* should today be bound together in this way to assemble the new, modern human *type* (and I use this term in the strictly Nietzschean sense) already determines much of what it is today possible to say and possible to know. For example, one would not expect to find many in the Western world, let alone in the microcosm of our own profession, who sincerely believes that the mastery of nature, or of chance in general, is *not* the best means for improving the plight of the human species on the planet. Our culture's seemingly unassailable belief in freedom and our faith in mastery are in fact twin and inseparable aspects of Western modernity, and together these form the almost theological basis and economic engine driving our technological civilization. Insofar as we are modern, technique, and its culture, is not a thing external to us that might be fixed, isolated, and examined; *it is what we are.*

For this reason among others, it is far easier to form in our minds the image of what our technological society has made possible for us, but in the wake of its successful and relentless onslaught, far, far more difficult to remember that of which we have been deprived. The very affective economies and systems of meaning against which such significant absences, deletions, and reconfigurations might yet be legible to us as historical aberrations are, in turn, the very ones that were targeted, broken up, and swept aside during the formation of the "modern" systematic mind. Mastery, at least our Western technological type, was achieved through techniques of quantification and numericalization, the reduction of shifting, variable fields of sensuous and material values to precise and fixable measurements. The full, if apparently crueler and more episodic world of pullulating qualities, of uncontrollable mixtures and the ecstatic values that both relate to and derive from embodied senses, once gave rise to different ecologies of both spiritual and social order, different grades of subjectivity of which today we are no longer able to summon even the faintest intimation. Modern society may be free of many older, systematic forms of political subjugation, much senseless religious and superstitious barbarism, and the random brutalities of chance, and it may be good to witness and celebrate the passing of these archaic forms into irretrievability, but that does not change the fact that every day and all around us, something is being lost.

Among the assertions I would like to put forward is the following one: In a society such as our own, because we are unable to separate discourses of emancipation from discourses of mastery, unable to dissociate the will to freedom from the application of technique, and because the historical fusion of these two modalities constitutes the birth and hallmark of our modernity, the very existence of a language of resistance to modernization may itself belong to the systematic but largely invisible series of deprivals and exclusions upon which our technological civilization is built.[2]

In other words, because the concept of abstract freedom is both the highest and most unchallengeable value in our modern world, having replaced virtually every other system of meaning both past and possible, secular and sacred, and because this concept of freedom is always already embedded within the boundless expanse of our belief in technological overcoming, no systematic critique of technological civilization is possible today in a language that would be intelligible to it.[3]

Is it possible, one might ask, that we suffer from a largely imperceptible, technologically induced collective type of historical amnesia, one that is entirely structural to our civilization? And that the peculiar distortions that such concepts as freedom, identity, and sexuality have taken on within our largely urban Western modernity have arisen expressly to hide from us this fact?

For among the most important developments occurring today, beyond the indisputable (and probably inevitable) media-technological revolutions that are reconfiguring our cities, homes, and workplaces, is one that may have to do not simply with the encroaching soft tyrannies and surreptitious deprivals that are an intimate and necessary complement to the so-called advances being prepared for us, but also with the withering away of even the language with which such continuing deprivals and subtle tyrannies might still be couched or expressed.

Jacquard Loom, 1801. The birth of code.

***

Though the threat once rhetorically posed to architecture by "the book" (in Victor Hugo's classic novel *The Hunchback of Notre Dame*) was in time revealed to be baseless, there does appear to be a challenge arising in our midst today that is at once far more systematic and comprehensive in its reformulation of classical notions of space and place, and far more spectacular in its capacity to invent architectural and environmental propositions for the structuring of life, labor, and communicative interaction. This boundaryless new medium of "virtual" reality is, however, *not* a simulated environment, as many continue to claim, but a new space altogether—indeed a new type of *total* environment, to borrow the adjective "total" from sociologist Erving Goffman, replete with its ominous association with prisons, mental hospitals, factories, and schools—a new *total* institution, made possible not by confining walls and political, social, and medical decrees, but by the seemingly "natural" evolutionary convergence of telephones, data banks, computers, and televisions. It is understandable then that at least some of the hype around the virtual world suggests that there is nothing threateningly new about it, that it is a full-fledged form of social life, only richer, more egalitarian, vaster and more free; and that its domain—cyberspace—is not unlike that of a modern city, only infinitely more capacious, complex, and delirious.

Cyberspace, of course, as the now classic adage goes, "is where we are when we are talking on the telephone." It is, in other words, neither in a *here* nor a *there*, but is a continual process of articulation, relentlesssly boring through us. In other words, it is not where "we" are at all, but where our *attention* is within a promiscuous, multidimensional, electromagnetic matrix, even when our bodies (once again bearing the brunt of Protestant-capitalist contempt) are hopelessly fixed in viscous Euclidean "real" space.

Attention, we must not forget, is at once effort and action; it is the application and distribution of energy on a disorganized material, a ceaseless feedback system that penetrates and arranges this material at a distance to produce a discernable pattern of distinctions and differences; it is a process that transforms raw time into history, giving it definable qualities, storable elements.

Wireless family, 1940s.

In a word, it is *work*, in both the thermodynamic sense and, alas, in the purely economic one as well. More than anything else, the virtual world may represent a landmark innovation in economic, as well as urban, architectural history—a new *ergonomic* interface in which the world becomes a huge, totally integrated factory/workplace, where the multiplicity of human being is once again forcibly reconfigured to isolate and affirm only those features of life and body that can be rendered productive—in this case, the magically complex and unreproducible *states* of its nervous system. With the help of any interactive platform device—video and audio headsets, gloves, wired bodysuits, or even a simple screen or computer terminal—one "inhabits" a virtual environment almost entirely through a stepped up form of the same mental or nervous vigilance used in navigating the flux of the real world. Only now, this type of vigilance, which controls—no matter how pleasurably—the interactions of vast arrays of sensual elements and relations, has become potential *productive* energy, and in doing so opens up vast new colonies for the production of surplus value. Listen to the virtual apologists incanting "Cyberspace is where your money is," thrilled that so much historical process can be compressed into so few familiar words, and nearly giddy at the thought that human affect and substance might finally be reduced to the same double-entry accounting practices as money itself. Indeed, the recent "realization" of the virtual domain does not simply introduce a new set of tools, or a new way of using tools, but may be heralding the advent of a new social regime in which all value and experience can be created and made intelligible only *in* and *through* a reality thoroughly saturated by tools.

***

Ought we not be concerned today with the apparent docility with which these intrusive and disruptive social, political, and industrial developments are being welcomed open-armed into the most intimate reaches of our domestic foyers and nervous systems? Is it alarmist to be troubled by the seemingly consensual chorus of enthrallment and slavish exuberance that has swept through our news media, our corporate and governmental boardrooms, and our college campuses whenever the promise—or specter—of a future "wired" universe is raised? I, for one, have no difficulty remembering a time when these three, very different social institutions constituted distinct and *vigorously* opposed mindsets and interests, nor have I forgotten the sting of

Book Party (for the party that hated books): House Republican leader Newt Gingrich opens a ceremony for the release of *Contract With America* in front of the Capitol, September 27, 1994.

disappointment and helplessness on witnessing their slow but inexorable migration toward and into one another during the Reagan 1980s, but it is no less shocking to behold the manner in which—in the midst of both the "beige" Democratic mood of the early Clinton era and the crude pseudo-Jeffersonian libertarianism that has now displaced it—both student and corporate executive, as well as politicians, professional purveyors of "truth," and culturati have come to share, almost without nuance, the same blandness of outlook and corporate habit of mind: Life in a machine.

Who could possibly have remained indifferent to the early 1990s news coverage of the incoming Newt Gingrich team's overhaul of ex-Speaker Tom Foley's old offices in the U.S. House of Representatives? I refer here to the unrestrained expression of derision on the part of newscasters as the cameras panned over what was presented as the Democrats' cumbersome and "outdated" electronic equipment, then over the typical rows and stacks of books that in any other time or context might have appeared reassuring, stately, refined, yet now, alongside the hard and real paraphernalia that visibly reflected the years and toil of governing, was dismissively displayed as the relics of a passé "paper administration" that had long since lost touch with the times. On the other side of the office—waiting—were the neat columns of the Gingrich team's factory-sealed, logo-stamped boxes of brand new computer equipment, scurrying teams of telephone repair people installing high-speed modem lines, and then, a closeup of the lone Republican bookshelf (three board feet!) with Gingrich's personal, streamlined, no-nonsense library composed of no more than six learned tomes—by the likes of Tom Peters, Peter Drucker, and Alvin Toffler. If anyone still managed to harbor any doubt of the type of barbarism being prepared for us, this single scene would have dispelled it definitively. When soon after, Gingrich, with the help of the right wing Progress and Freedom Foundation, issued his unscrupulous "Magna Carta for the Information Age," the golden—if deeply naive—era of cyberspace frontiership probably came to a close. In other words, the new gold rush had begun (the call for universal deregulation and the repeal of so-called "Second Wave" legislation and rights), and along with it, the corresponding vision of a profoundly segregated new society (outlined in the promotional video "Renewing American Civilization"). By 1995, exactly twice as many Internet users in America identified themselves with Republicans (48%) than with Democrats (24%).[4]

Clearly, the idea of a fully synthetic, infinitely reprogrammable world where even our own identities may continually be shifted and transformed offers unbounded possibilities to the speculative, political imagination. But do virtual reality, multimedia, and interactive telecommunications really represent just an extraordinary new space of limitless play, social intimacy, and invention, or is the logic behind its imminent emergence more systematic, more historical, more nefarious, and overdetermined than its apologists let on?

To peruse the travel brochures for the new information highways—or the "infobahns," as they are ominously often called5—is to come across exclamations of the following type: "The Digital Revolution is whipping through our lives like a Bengali typhoon [provoking] changes so profound their only parallel is probably the discovery of fire"; this from *Wired* magazine, once the English language world's hippest con, and marketer of the new corporate-youth ethic. It is to be led to believe that we are on the threshold of one of the great political notions of the Ages, one endowed with the capacity to foreclose existing dialectics, to bring the great chain of being to its full expression and final resting place, to reconcile within acceptable parameters outstanding differences—political, ethical, sexual and economic—between humans. To survey the political literature on cyberspace is to be led to believe that the medium's gravest outstanding problems concern only those of regulation and control, never the simple, brute fact of what is arguably *an unbridled invasion and seizure of power of the public sphere, of the modalities of the human perceptual apparatus, of human energy, and of the interdependent historical ecologies that together these three entail.*

***

It is probably true that since the early nineteenth century urban transformation has been driven much less by building innovations and architectural schemas than by the cruder, uglier, but irresistible forces composed and unleashed by the evolution of machines. Yet already in the early Renaissance, flexible, rational technologies of dissociation, such as the clock, the printed book, movable painting panels, navigational instruments, and modern accounting practices had initiated what could later only become a massive wave of "deterritorialization," that is, the disentanglement and analytical isolation of previously embedded values and flows. More than anything else, the techniques that issued from these early modern technologies permitted the rise and growth of the first "artificial" cities, for the simple reason that abstract economies based on trade and ledger sheets could for the first time be cultivated entirely independent of constraints of space and real, local productive capacity.

By the nineteenth century the development of lithographic techniques allowed images to follow text into the public domain of mass reproducibility and mass circulation, while the emergence of the literary form of the novel, as well as the newspaper and later of magazines not only generalized literacy through the emerging merchant middle classes, but invested the city proper with a whole new *informational* superstratum or network, complete with the system of exploitable class structures and exclusions that such informational economies invariably entail: access to books and newspapers, after all, was never meant to furnish access to literacy; rather, the reverse.

The new printing techniques gave rise to a whole new capitalist culture of romance consumables, and in turn, new economic megastructures such as the department store, thanks to the advent of advertising culture and its consequent recoding of urban social space as a marketplace continuous in space and time. The manufactures, the railway stations, and, undoubtedly, from a more distant background, the inhuman mines, fill out the schema of the nineteenth-century urban landscape, each environment in its own direct or indirect way made possible by combustion energy and the steam engine.

By the turn of the twentieth century there emerged the modernist city, reshaped and considerably cleaned up in accordance with the new reigning principles of bureaucratic organization; this is a civilization built increasingly on the concept of files and their administration, a culture of records and microinformation, of rationalized management, of hygiene and scientific production. Urbanization increasingly begins now to follow the lines of managed density and increasingly subtle, enriched forms of mass dissemination that include the first electromagnetic technologies such as the cinema, the telephone, the wireless radio, incandescent light, automobiles, and the eruption of a third, vertical urban dimension thanks to the elevator, airships, and airplanes.

From the early nineteenth century till the second World War, remarkable urban visions came and went—Dickens and Engels in London and Manchester, Haussmann in Paris, Leonidov in Moscow, Ferris in New York, Sant'Elia in Milan—but the ever-transforming informational economies of cities did not win definitive success in breaking the morphological barrier of the classical capitalist city until the end of World War II. Suddenly, the automobile, combined now with television—and only in such synergistic combination—induced a new type of urbanization: scientific in the extreme and totally reliant on wartime engineering structures and practices, the city explodes spatially, but only as a quilted interlock of increasingly confined and abstract synthetic environments. The chilling presence of engineering could be discerned virtually everywhere, controlling not only the spaces but even their syntax of use: from the freeways, to the office buildings, to the suburbs. Munitions factories and the full-employment war economy of which they were a part are now also turned inside out: the women who labored in them are now the subject of the most intensive technological reengineering putsch of all time as the new workplace concept of rational, or numerical, ergonomics is projected onto the domestic household and backed up by a hard-sold panoply of so-called "labor-saving" domestic appliances. Reality is increasingly carved up, processed, and filtered, then repackaged and delivered in controlled envelopes of remarkable persuasiveness, primarily because *there is no longer anything more real (less engineered) against which such synthetic constructs can be checked.* Clearly television also played a determining role, not only by replacing contact with the real multispectral information fluxes of the true city that had been traded off for the pseudo-pastorality of the suburb, but because the reality it delivered was itself totally engineered, oriented to merchandising and thereby to extending the ethos of the mastered environment to ever finer levels. In many ways all other social and economic developments concerned mere logistics and matériel, but television arguably comprised the veritable nervous system of the post-World War II world; its role was not just to compensate (for the quick of the city and the marketplace that was lost) and to deliver (entertainment, information, goods), *but in effect, to retrain.*[6]

***

The first—and most difficult—thing to understand about television is that it does not represent an isolated object in a landscape with a circumscribed field of effects, but in fact comprises a full-fledged environment of its own. Though today we in the West receive far more information than at any other point in human history, we certainly receive far less than ever before in an unfiltered, raw, or unmediated form, in what might be called its "whole" state, that is, naturally embedded in a sensuous, complex array and apprehended directly by actual experience. Indeed, information and experience and their difficult relationship (they are, in effect, for the most part, opposed to one another) together comprise the new battlefronts to be seized and occupied by technologization, capitalization, and power, though in a way decisively different from earlier historical regimes of subjection. This is because the passage from the management and coercion of physical relations and bodies to the imposition of precise patterns and habits onto the fluid, labile continuum of human nervous response, represents a movement from the crudely empirical and visible to the subtle and invisible. Our modern, designed environments, including an ever-increasing proportion of the real wilderness that makes up our most vital cities, is already a restricted, filtered, *managed* reality where the predigestion and selection of data flow is both the very reason of their existence and their main, and most proud, claim to efficiency. All background noise and all free, non-channeled flows are eliminated in the name of creating frictionless, "dedicated," or task-directed environments. Such environments work primarily either because they are designed expressly for the efficient deployment of other, smaller, similarly reduced machines (imagine anything from a freeway with its automobile machines, to a laundromat with its washing and drying machines), or with a view to effectively channel human attention, to control it, manage it, and direct it into productive but limited "activity pathways" in an increasingly optimized way (any "well-designed" modern office or factory).[7] And yet, it may be said that no inert spatial environment can successfully engage the mobility and fluidity of human attention as fully and assuredly as a dynamically active electromagnetic one.

Our nervous systems, evolved over millions of years in predator/prey environments that favored massively intertwined, experiential modalities over discrete, abstract, informational ones, naturally read movement and temporal processes more readily and with more acute and intense engagement than fixed, purely spatial ones. In other words, "This (television)" may well already "have killed that (architecture)," to reinvoke Victor Hugo's famous prediction. Herein lies the specific beauty and insidiousness of television: it easily lures attention with its flow—television lures attention easily because it lures it initially along its pathway of least work or resistance—then, once captured, confines, and tunnels attention into rigid, disconnected pathways of predigested, continually—but only infinitesimally—varying monotony. One would be hard-pressed today to imagine any advanced, designed environment to which this latter description does not apply as accurately as it does to television.

Consider, for example, the ubiquitous modern shopping malls built since the 1960s. Mall engineers long ago identified a type of environmentally induced anomie or vertigo—a kind of queasiness and disorientation identical to that experienced by many TV watchers—that became known to the planning industry as the "Gruen transfer." This term refers to a threshold, the moment when a shopper's purposive behavior and directed, coherent bodily movements break down under the barrage of excessive, narrow-spectrum stimulation and continual interruption of attention. The unconsciously bewildered shopper, rendered docile, cannot help but drift into the prepared pathways and patterns of externally induced consumer activity, unfocused but exquisitely suggestible to gentle but firm environmental cues. It is estimated that as much as 90 percent of shopping mall transactions concern goods that the shopper had not planned in advance to buy, yet did so after the Gruen transfer took hold.

This returns us to the problem of retraining. The human organism is unarguably a marvelously flexible entity with enormous range and capacity to integrate even extremely subtle changes of features in the environment. However, as the milieus in which we live become more and more channeled, rationalized, efficiency-oriented and engineered, human capacities—both to affect and to be affected by the world—become filtered, narrowed, or truncated. Every "rational" environment is in fact a *machinic* environment oriented to the efficient use, or flow, of energy and material resources and their conversion into value, if not explicitly into goods. Early synthetic environments such as the factories of the Industrial Age were not only sites of concentration of capital and production, they were also sites where the actual *discipline of labor* was first forcibly imposed on the human organism in such a way that the nuanced, almost infinite multiplicity of the human motor-nervous response could come literally to be subsumed by the primitive, sterile, repetitive modalities of industrial machinery. These factories, or "labor locales" were, quite simply, training grounds: that is, in addition to serving as sites for production of surplus value, they operated as sites for the production and formation of an altogether new "type" of human being. Yet, as the human relationship to machines became ever more intimate and scientific, for example, in the hyper-rationalized Taylorist and Fordist factories of the early twentieth century, it could be said that *the muscle-work of the pre-industrial age, had almost completely given way to the new, more refined, and invasive reality of nerve-work.*

In this period, and in the decades that followed, the concept of "management" becomes the organizing concept in the science of work. Management principally targets and extolls *integration*, the integration now of a whole group of submodalities that indiscriminately includes machines, markets, labor, and goods. With the emergence of this new, soft mega-entity, the human "unit," as it is now called, is analytically broken down, so that "routines" may be individually extracted from it and selectively integrated into this or that point in the production process; the human becomes a kind of universal servomechanism essentially monitoring and controlling—by means of its unique capacities of *attention*—an infinitely entailed cyclical process that seems otherwise to run on its own. Our machines may now well have their own independent personal and social lives and perhaps their own motive force, but they are, in relation to the massively complex, evolutionarily interwoven systems of the natural world, still remarkably crude algorithmic assemblages, and so they must, at almost each stage of their operation, parasitically tap the human nervous system to keep themselves up and running and smoothly coordinated. What the majority of humans do nowadays—for a living, as it were—is essentially *watch*, *discern*, *correct*, and *respond*, and it is these sets of nervous processes, and others closely related to them, that our labor and management theorists call "work," and that our credulous techno-enthusiasts call "communication."

***

It was hardly a surprise, therefore, that between mid-1993 and 1994 the big news in white-collar civilization was that telephone companies had started buying cable TV companies, who in turn were buying movie companies who were buying computer companies who were themselves buying telephone companies—and yet, in point of fact, it was not companies at all, but rather *entire industries* that had begun to converge in a rapacious territory-grabbing dance around the domestic TV set, spawning the largest corporate mergers in history. What emerged after the deal-making frenzy—despite the ultimate failure of certain well-publicized mega-mergers—was an information industry oligopoly of unprecedented scale and scope, where so much control over so much of what in the future we will call "culture" had been concentrated in the hands of so few interests that the sudden proliferation of hubristic phrases such as "the digiverse," "the telecosm," and the future "communicopia," could no longer really provoke laughter at all, but only a slowly escalating anxiety.

The new Virtual Society and the cyberspatial, multimedia systems which form its infrastructure are largely being sold to us as visionary, romantic, and hyperbolic expressions of many fascinating, even if less spectacular, technical developments, either already in process or just around the corner: universal digitalization and integration, intelligent systems, flexible specialization in the manufacturing sector, piezoelectric and microprocessor controlled materials and architectures, and the multiple revolutions in man-machine interfaces, which will destroy any preconceived notion of what the coupled words *man* and *machine* individually once meant.

***

This is clearly no garden variety, vicious corporate assault in the classical sense: the campaign of persuasion-coercion has come to permeate our society from all sides. The slickest approach perhaps comes from credentialed countercultural frontiersmen (they are almost uniformly men) who publicly pose as committed hackers/crackers and liberators of privatized information (a position that is supposed not to be vitiated by their stockholdings and memberships on boards of trustees for many hard- and software startup corporations). They argue that the technical infrastructure of the new virtual communities—despite its well-known origins in geo-strategic, military, and industrial applications—will create decentralized, highly intimate environments of human collectivity and free communication, and especially will dismantle the informational autocracies erected in the 1970s and 1980s during the computer and information revolutions (these latter, of course, have now suddenly become dismissably paleolithic). These cowboys are so gung-ho are on empowering "the people" that it does not seem ever to have occurred to them, despite the historical, demographic, and intellectual developments of the last two decades, that "people" might also mean workers, women, and those economically, and perhaps permanently, disenfranchised through race, culture, or geographical situation.

In fact, there are many obfuscations here. Among the most basic is the very supposition that information is even pleasing, let alone "liberating." Isn't information among the ugliest, most debilitating, and distracting nuisances in our lives? Here is the voice of one of the early, big playing sellers, Arno Penzias, Nobelist and vice president of research at Bell Labs:

> People are frightened at the thought of getting too much information, which just shows we're not in the Age of Information Transparency yet. Are you frightened by the thought of getting too much money? Too much happiness?. . . And when will you know you're in the Age of Information Transparency? I will tell you. If somebody says I can get you ten times as much information as you have now, if that makes you feel good, you're in the Age of Information Transparency. And until then we're in the Age of Paper Work.

Just what could this transparency mean? Who would want to be so transparent that she or he would spend ten times more of her/his day in front of a CRT screen, ten times more time searching data bases, ten times more time reading vast, unedited expanses of shapeless prose, ten times more time logging on, sending thanks, uploading, downloading, file transferring, scanning, etc.?

VORTEX

Silicon Valley aerial reconnaissance flight, Bureau of Inverse Technology, 1997. These images were taken from a 32-inch wingspan radio-controlled airplane that was able to fly un detected into the no-photography zones above Lockheed Missiles and Space Company, Hewlett Packard, and Sun Microsystems.

Of course Penzias is probably just referring to ten times more access, tantamount to, and about as useful as, for example, owning an enormous ranch that one *could* use, that is, if one didn't also already have houses in Paris, Aspen, Tokyo, New York, and the Amalfi Coast. Given that our time on earth is finite, having access to a database of 800 encyclopaedia rather than to a single good set can at best improve our effectiveness only incrementally, though it will certainly degrade the overall signal-to-noise ratio emitted by our world to almost useless levels. And yet, if this is what the information revolution offers *us*, one should not be surprised if what "they" stand to reap from our participation is of considerably greater value and interest. For we all know, or at least suspect, that while more information is terrible for life—it cannot not make an afternoon on a café terrace better, a smile sexier, or an idea more startling; in fact, it only impoverishes these by interrupting them[8]—it *is* good for business, though primarily through the systematic invasion of public privacy by rendering all of our actions visible, trackable, systematizable, and capitalizable.

Consider the implications of computer-based automation, for example, on "smart" roadways, letting aside for the moment the far more comprehensive colonizations which would characterize the virtual electronic superhighways:

Imagine some future government, faced with paralyzing traffic congestion in its urban centers and on its highways, declaring roadways to be private right-of-ways, and installing a program requiring motorists to pay fees for the privilege of using them. Since the idea is for motorists to pay more for a given road during peak congested hours, less during off-hours, but to pay in proportion to how much time is actually spent on them, and in relation to what designated zones were occupied—premium, subpremium or non-premium ones—a relatively intricate electronic toll system is necessary. Such a system would also have the capability to weigh freight trucks electronically (without requiring them to stop and pull into a weigh station), to check the weight against the truck's bill of lading, and then to debit the operator, all automatically. In addition, such a smart system could, and might even want, to monitor the speed of a particular Ferrari every 1,000 yards (which is excellent business), while simultaneously keeping an accurate tally of each time a particular Toyota runs an orange light or parks illegally, even for only a few minutes.

There are, even in our hyper-technical era, very few architectural works of foundational radicality or innovation. Frank Lloyd Wright's Guggenheim Museum in New York is one such building that somehow succeeded in breaking fully from the unbroken tectonic conventions of two millennia. The Guggenheim in New York stands up almost as if by kinetic expansion alone, without the assistance—or the reassuring lexicon—of posts, walls, and lintel-like supportive integers. (How different, and deficient, is Frank Gehry's Guggenheim Bilbao by comparison.) Clearly it is not every day, nor even every époque, that succeeds in producing a totally new model of large-scale assembly.
Yet, in our era, Toyo Ito's Sendai Mediatheque is one such époque-shattering building.
And, what is more, it is linked to the earlier Wright building by an important cosmological constant: the vortex.

The Sendai Mediatheque is visibly organized around a series of swaying, helical tubes loaded with rotational forces that permit extraordinary acrobatic departures from Newtonian convention. The global solid that is produced by the planes that are arrayed around them is virtually an *atmosphere* sustained by pressure differential, rather than a load-bearing member (as the pressure varies the "warp and the weft" of the plates syncopate against one another). The system of reticulation that is so visible and prominent in the twisting vertical tubes is thus repeated in rarefied form in the rectilinear plates that form the work's more commonplace use-surfaces. But this does not diminish—in fact it augments—the building's total "Gestalt" effect of presenting itself as an overall textilic or woven solid. The Sendai Mediatheque induces a world of floating and drifting elements, semi-emancipated from gravity, by espousing one of the earliest forms of material technology: what one prehistorian referred to as the class of "supple solids" best typified by the early basketry arts. The mastery of supple solids represented a radical moment in the elaboration of early human technology and the domination of one part of the physical and material environment. Today, advanced geometry, mathematics, alloys, and "smart" or responsive matter—and in conjunction with some deeply original thinking—may well be opening material speculation to solutions that even five or ten years ago were unimaginable.

Toyo Ito's building in no small way appears to have introduced a new paradigm into architectural thought. The nest-like structure of the new Herzog & de Meuron and Ai Wei Wei-designed Olympic Stadium in Beijing, to name but one widely-noted structure, as well as the host of mesh-support structures on drawing boards everywhere today, are in fact homages to the groundbreaking audacity of the Sendai Mediatheque.

ダイアパレス
春日町

Magnetic induction loops buried in the roadway would act as the system's sensors, and they could, in addition to everything else, time and coordinate traffic lights, count vehicles, monitor traffic speed, and feed all this information back—with rerouting advice when necessary—to the user-drivers via roadway billboards, or more directly, to small dashboard-mounted screens inside the vehicles. More advanced roadside or satellite sensors of the bar-code-reading variety will obtain much more detailed information from a vehicle's electronic tag (Automated Vehicle ID or "AVI"), a technology borrowed from military aircraft "friend-or-foe identification" systems.[9] Drivers in turn would either carry their AVI tags in a conspicuous place, allowing roadside or airborne radios to identify the car and automatically to deduct the toll from the driver's AVI account, or better, they could purchase "smart cards" (like those inserted into telephones to direct billing to a home address) that could be swiped through a device inside any vehicle, and then, via the AVI tag, be billed personally for whatever tolls they racked up. An elegant system.

Now the downside: the latter scenario was not drawn from a science fiction novel at all. Such road systems already exist, in limited fashion, in Belgium, France, Japan, Germany, Sweden, Singapore, on America's interstates throughout the Northeast, in London, Newcastle, Cambridge, and Oslo—indeed, they are poised to become more or less universal.[10] But the second, and far more serious, downside is that such smart road systems will, once deployed, have the capacity to know precisely where a motorist *is* in the network at any given time, in fact, where they had been, when they were there, and for exactly how long. Indeed, such roads will almost certainly know far more than this, depending on which traffic-management and road-pricing schemes are ultimately implemented. This electronic tollbooth urbanism is little more than one massive, automated traffic RoboCop nightmare in the making. Combined with both traditional and video cameras, image-processing computers and databanks, the system can easily be (and already is) designed to nab drivers without cards, with outstanding tickets, lapsed inspections, or who simply have bad credit. The day will all too soon come when we will need to ask exactly how much are we willing to allow our roads to know about us.

Things will only be worse—to the nth power—on the *information* superhighway, for the simple reason that far more intelligence is being built into the system, both at the terminals and in the networks themselves. In fact, bank, credit card, and telephone use have always generated a frightening amount of information about our lives—try examining a friend's or lover's phone bill—yet in the approaching era of the wireless "online" citizen dutifully (as if it were truly a choice) toting a "personal digital assistant" on the wrist or in the back pocket, the integral "police-function" that Jeremy Bentham handed down to us in his efficient prison designs from the eighteenth century will have been permitted so thoroughly to suffuse our daily reality that an offline jail cell may one day seem like a rare, limited space of pure freedom.

Add to this the increasing prevalence of Automatic Number Identification (ANI) systems and Caller ID, which are already sending not only our names, phone numbers, and billing addresses but now also database files, credit information, and consumer records to any business we call using an 800 or 900 number, and this even before our voice connection is even accepted, not to mention the compiling, analyzing, and selling of this data to other marketing firms on the model of mailing lists. AT&T has already whorishly honored their commitment to switch exclusively to using the U.S. Government's Clipper encryption chip on its networks, granting federal eavesdroppers master-key privileges not only for wiretapping but for monitoring all telecomputer traffic at will.[11]

Yet, as I indicated earlier, thanks to electronic footprinting, the networks themselves will know far more about us than any centralized government ever will or could. Naturally, hundreds of irresistible new services are being developed for cellular communications systems, most based on the necessary premise that we users will let the switching network know where we are. Most will use global positioning systems derived from maritime and military satellite-based devices.[12] Marketers will map our walking routes and displacement patterns in a city with comprehensive precision, and though this information will most likely be used benignly to locate sites for new Burger Kings and Federal Express outlets, on the other hand, maybe you just don't want someone else to know that you're always taking that trip at lunch.[13]

Just as advanced television technology ultimately came to be used in the 1980s for surveillance applications, it appears that the phone-based, telecommunicational digiverse just may be on the verge of delivering the real, and terrifying, total transparency that Arno Penzias once talked about.[14]

Despite all this, we continue to hear daily a chorus of what marvelous new freedoms and socialities await us on the networks: how we have the choice to be anyone we like, how we have the freedom to crossdress, become plural, multiply ourselves, become old or young, talk dirty, act out any sociopathic experiment we like without hurting others or humiliating ourselves too badly. This may be true . . . for now. But as interfaces mature, the dream of mapping our bodily responses directly and seamlessly into cyberspace will become increasingly seductive and real. Current research and prototypes are based on ultrasound and magnetic resonance technologies, as well as interfaces derived directly from lie detector technologies; they monitor body temperature, blood pressure, heart rate, as well as micromotor responses in such a way as to generate massively detailed, personal body response profiles—indelible, unchangeable signatures—that no crossdressing, role-playing, or Halloween mask will be able to mitigate.

Clearly, emigration into these new environments is not only today being made quasi-mandatory through public intimidation programs about progress, global competitiveness, and how little Johnny and Sue will be left on the dole lines if they don't get wired by age four, but it is certainly full of risks, impoverishments, and illusions. Access to the nets is not now, nor will it ever be, equally and democratically distributed throughout society.[15] Elitist forms of urbanization are already standard fare with expensive "country club" environments and services in which the well-to-do connect and network efficiently among themselves.[16] The entire Internet universe might itself be conceived as a kind of blueprint, belonging to the classic *deurbanization* model of a wholesale secession of a unified social class from a public sphere to a protected enclave (the class referred to here would be one of the new informational classes such as "those who live or work in synthetic computer environments").[17]

In fact, the idea of unification and preselection may be one of the worst dangers shaping Internet sociality. Because the Internet is an entirely engineered space adjacent to, and entirely removed from, the concrete material world, the normal constraints of friction, viscosity, noise, and the general panoply of resistance and reactivity that characterize responses to gestures in the material world are nearly absent from it. For this reason, aggregation patterns, unlike those traditional ones unfolding in richly articulated, complex, material social spaces, here *freefall*, in the absence of cohesional counterforces, into micro-user groups of unbelievable, even shocking narrowness.

This narrowness is then in turn extolled as some type of "revolution" that, among other things, will soon allow the preselected, predigested "narrowcasting" of personalized daily information in a neat package that will replace the daily newpaper with what MIT Media Lab founder Nicholas Negroponte calls without irony "The Daily Me." More social and intellectual atomization, just what is needed for today's fragmented, creolized, ghettoized world in crisis.

Most important of all, with the migration of tools from the older, peripheral spaces of human ecology to its center, and with their transformation from an external, objectifiable object into an *environment* and a set of modalities that are inhabited, human substance and the capacity *to conceive* of human freedom are modified in the process.

Somehow, somewhere, our democratic institutions will need to be reinvented, if only to lay down the rules and fix the limits within which economic rationality ought to be permitted to function and to run roughshod over our lives. The forces behind the coming virtual city are driven by savage economic imperatives, not delirious new possibilities for freedom, and we will need, at great challenge to the imagination, to find effective new ways of refusing them. We will need to create strong alternative cultures resistant to the sleazy short-term seductions of gadgetry, in the hopes of maintaining any semblance of longterm autonomy over our fates. "Jacking in" to a cyberspatial matrix, as the characters in William Gibson's foundational novel *Neuromancer* put it, is, after all, hardly a convincing idea of life reinvented beyond the tyranny of productivist ethics nor the most promising erotic vision of a world where the body's energies are partially freed to create new modalities of pleasure.

But in seeking to develop the new intensities, the alternative, offline forms of culture and attention, we will almost surely have to free ourselves of certain prejudices that once served us well. Indeed, the most difficult part of all is the possibility that for the first time in our history, and against all of the truisms of our education and our traditions of progress, it may well at last be necessary, while cautiously stepping forward across the ominous threshold into the future, to keep at least one eye hard-trained on the past, on its *objects*, and on the parade of *types* that make up the humans that we once were. Somewhere in that pageantry a million little spores lie, each representing a pathway once primed but never taken, along which human consciousness might today still travel and free itself from the precariously narrowing trajectory on which it is otherwise perilously embarked today.

1 What is presented in the following pages makes no claim to be a work of scholarship. It should rather be understood as a *performance*, a manifesto to the radical acts of private and collective memory of which we are all capable, and to which we must once again—without apology—find our way.
2 I would like to acknowledge the direct debt that these ideas owe to the work of two philosophers of technology, George Grant and Jacques Ellul.
3 The so-called Islamic revolution of the last twenty-five years is but one example. Beyond the neo-archaic, blindly anti-American belligerence, it represents a broader, more systematic—and for these very reasons, (to us) nearly inapprehensible—refusal of what has become the Western ethos of modernization, secularization, and technologization.
4 *Newsweek*, February 27, 1995. Even the sheer numbers of Internet users that are so often used to promote the new services and habits are possibly only one tenth as high as commonly reported. Cf. *The New York Times*, August 4, 1994, p. D1. For a more sustained elaboration of the importance of the Gingrich doctrine on American Urban civilization, see my "Contract with America," *Mutations* (Barcelona: Actar, 2001), pp. 566–601.
5 The German *Autobahnen* were developed to facilitate the logistical and supply requirements of the Nazi war machine during what would become the second World War II.
6 See my "Television: the Infrastructural Revolution," *Mutations*, pp. 508–519.
7 Antonio Gramsci, the first philosopher to describe social relations in terms of muscular-nervous effort, argued that one could gauge the complexity of industrial societies by their capacity to produce machines with which to produce further machines, and that of technical civilizations by their capacity to produce not only scientific instruments but also to produce further, more refined instruments for testing the first ones. In this same sense, our own society moves toward ever greater mathematical and perceptual abstraction as it engages itself in planning and engineering environments specifically for engineered—and no longer primarily natural or animal—objects and flows.
8 "People don't even know how to want a lot of the stuff we're going to offer," claims one manager at Microsoft Corporation. "Survey on Television," *The Economist*, February 12, 1994, p. 10.
9 First used (with less than impeccable accuracy) in the First Gulf War in 1991.
10 *The Economist*, August 7, 1993, pp. 71–2.
11 The National Security Agency (NSA) has reportedly monitored all international phone traffic in the U.S. for decades—all incoming and outgoing calls, as well as all those simply transiting through the country—based on machinic recognition of targeted keywords. The new key-

stroke-based communications and optical form- and character-recognition technologies now not only increase the detail and accuracy of these surveillance techniques but extend their domain of applicability by an exponential factor. The advent of the Homeland Security Enterprise has, for the time being, removed virtually every barrier to these developments.

12 A new "follow me anywhere" lifetime phone number system, with so-called "area" code 500, was approved by the FCC and has been available for over a year. It tracks customers anywhere on earth and forwards calls through existing infrastructure. Very soon new, satellite-based systems will be deployed by companies such as Iridium-Motorola, Qualcomm, Odyssey-TRW, a defense contractor, and Inmarsat, a company that formerly provided satellite services to sailors. Cf. *The Economist*, February 11, 1995, p. 73.

13 James Gleick, "The Telephone Transformed—Into Almost Everything," *New York Times Magazine*, May 16, 1993.

14 In what is considered to be a successful pilot project, more than 300 British streets have been wired for 24-hour surveillance by closed-circuit police television cameras (the "securicams" in Max Headroom's 4,000 channel universe?) integrated into existing streetlight infrastructure. A version of this system will be used to scan crowds in Atlanta at the 1996 Olympic games. Built-in intelligence will allow individual images to be relayed for identification. Cf. "Rights of Privacy," *Scientific American*, November 1995. Costumed cartoon characters at Disney parks have long been equipped with electronic eavesdropping and recording devices to police crowds.

15 "Survey on Communications," *The Economist*, October 23, 1993, pp. 4, 15. At the time of this writing, fewer than forty-five percent of Americans have access to their own phone lines, less than one percent of Chinese and Nigerians, and only about one in ten people on earth.

16 On the complacent role of contemporary criticism with respect to these questions, see Jonathan Crary, "Critical Reflections," *Artforum*, February 1994. Universal service laws in the United States are already hard-pressed to protect the rights of the disenfranchised and the marginal in an unregulated Republican-model market. It is worth pointing out that the average American household already pays almost $500 a year in embedded costs for its so-called "free" television service.

17 Consider as well Nicholas Negroponte's brutal coinage, "the digitally homeless," and his less than ingenuous speculations on the future digital democracy in his monthly column in *Wired* magazine (of which he is, incidentally, co-founder and -owner). According to his friends, "90 to 95% of Americans" will have e-mail addresses within four years. And how many will have real addresses?

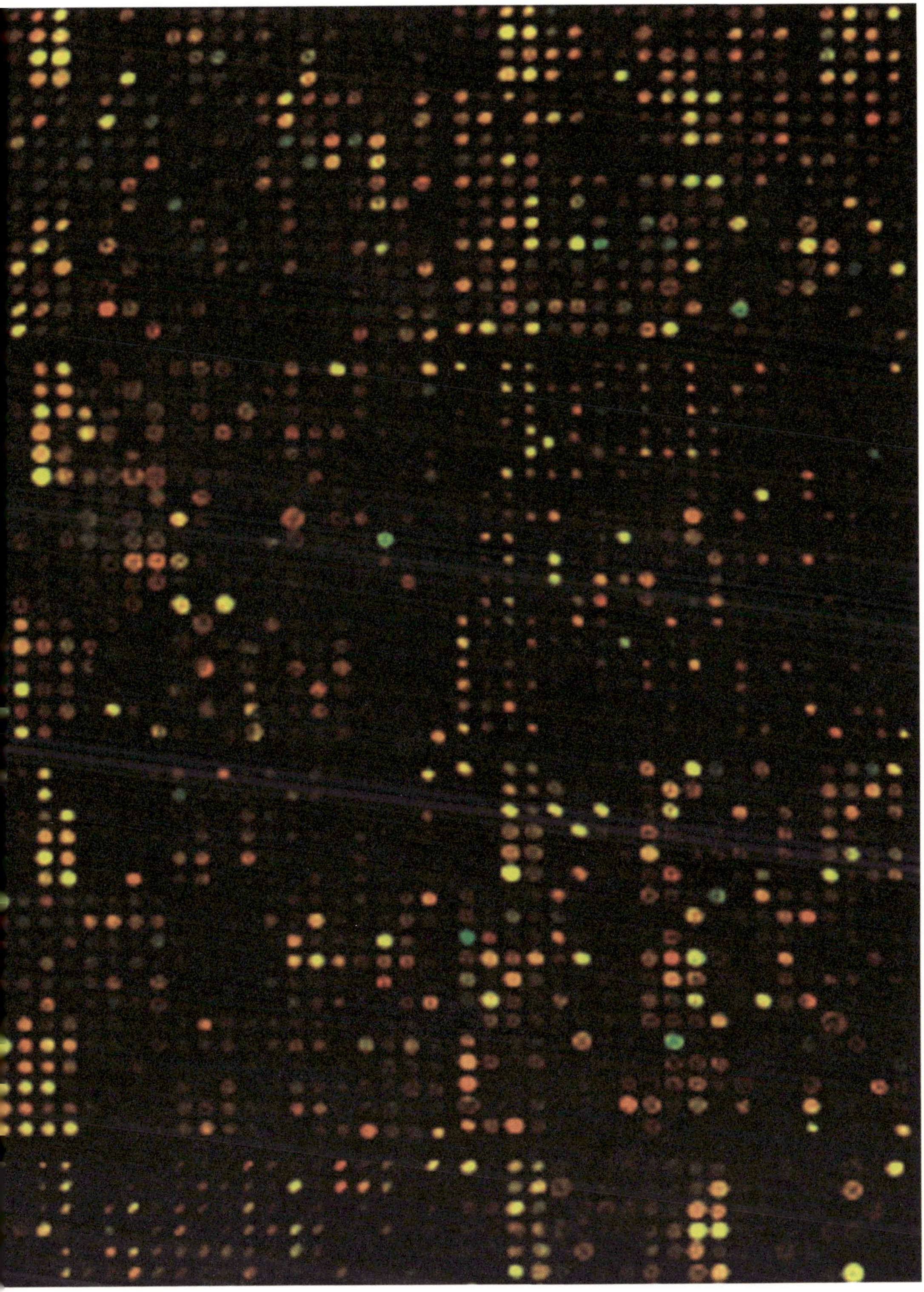

Human Genome.

Few people would disagree that our civilization, our ethics and our politics are through and through technological ones. And yet the problem lies not here, but rather in the fact that we as beings, as animals, as examples of a once human *nature*, are so fully and irretrievably technological ourselves that we are today deprived even of a language with which to articulate, even to summon a memory of, the historical passage toward this fate which has claimed us. The Western technological imperative—the will to mastery of an indifferent nature and of the inadvertent cruelties of chance—simply cannot be thought separately from the more general (and however apparently noble) dream, of the material and spiritual emancipation of humankind. Since the Renaissance, "Man" has defined himself to be—in his essence—a creative *freedom*, and this ideology is the source both of that liberal humanism only lately maligned and fallen into disrepute, and the indomitable and more robust drive to realize this freedom by conquering and subjugating nature through technique. It is for this reason that it may be still necessary to think of our technological being, our condition, as *a fate*—for we are simply no longer the type of human we once were, and we are today scarcely able to imagine what other type we might become, if not in relation to this all-too-modern "type" of embodied freedom.[1]

Until the nineteenth century, Human technics in the social, political, scientific, industrial, and domestic domains, was overwhelmingly oriented to the mastery of adversity and chance in the external world of nature. Since the turn of the nineteenth century, however, the will to mastery has been at least partially turned away from the non-human world and applied to the mastery, knowledge, and control of a *human nature* as well. The rise and development during this period of the knowledge industries known today as the social or human sciences were little more than logical—that is, ostensibly humane and cultivated—responses to a civilization whose technological ethos has become no less than its prime motor, and for which political administration and enforcement, as well as the control, planning and integration of human populations into this ethos, has become paramount. This new stage in the development of western technics was, in any case, far less a response to a real, philosophically or organically formulated human need, than a response to

ARCHITECTURE AND THE TECHNOLOGIES OF LIFE

a crisis (social and ergonomic) generated by the irresistible march of technics itself. Since that time, technology has been oriented more and more, and with increasing, even terrifying refinement, to the control, management, and now *planning* of human activity.

What is crucial in these developments, though also powerfully hidden by them, is that the processes of *making* in our culture, and the processes of *knowing*, are no longer separate. Only through the genealogical methods of philosophy (and through its nearly outdated modalities of attention and concentration) can this fact still, however dimly, be glimpsed. Nor is it by any means inauspicious, nor unimportant, that the problem of technology is increasingly being introduced today into the rich intersection between philosophy and architecture. And this, not only because *knowledge* can be said to be the proper domain of philosophy, and *making* the proper domain of architecture, but also, because it is the peculiar, privileged, and possibly insidious ways that architecture envelops the technological ethos through its own modalities of knowing, and the ways, both hidden and still to be invented, that philosophy may yet function for us as a crucial new modality of *making*, that is at stake today, and in the decades to come.

The fact that our technological civilization, since the nineteenth century, has turned its "knowing/making" attention ever more acutely toward human and other organic milieus—to the study of society, the "psyche," history and economics, as well as the tissues, molecules, and processes of life—is among the most important developments of our modern history. In 1934, in a magisterial work on the history of human technics, Lewis Mumford outlined in broad strokes three great periods of western technological development: first, the Eotechnic Period, characterized by the technologies of water and wind power, glassworks and wood; then the industrial or Paleotechnic Period of the dark mines, crowded factories, and blackened cities, of giant ironworks and machines, and energy derived from combustion and pressure, steam and coal; finally the Neotechnic Period, from the midst of which Mumford himself was then writing, a period whose innovation was to have hitched its *techné* irreversibly to the precise and minute operations of mathematics and the seemingly infinite fecundity of science.

First presented as an address to a conference at Columbia University, November 1991.

Nanotube assembly line. Nanotubes, made of cylindrical fullerenes, the material that brought us "buckyballs" in the 1990s, are the first totally engineered molecular structures. One of the strongest and most conductive materials on earth, this pure carbon matrix has found uses in combat jackets, concrete, bicycles, computer circuits and artificial muscles.

This last period delivered the technics of alloys, light metals, synthetic compounds, the invisible mysteries of chemical processes, and especially electrical energy, as well as reintegrating long neglected "wet" milieus such as the farm, the vineyard, and the physiological laboratory. Yet near the end of his book *Technics and Civilization*, and with the suddenness and singularity of a geological eruption, Mumford marks, if only in passing, the chilling apprehension of what he calls "the biotechnic period, already visible over the edge of the horizon."

There can be no doubt that Mumford's almost impossibly prescient vision is now emerging with a force at once as awesome and terrible as it is ineluctable. The unprecedentedly huge program of global administration in the name of the AIDS epidemiology project, and its correspondingly massive immunological and pharmaceutical initiatives, the almost anarchic rise of the new biology industries as if these were little more than structural refinements within the familiar Silicon Valley paradigm, and the Human Genome Project that subtends them, a public works project so colossal that it is fully military in its proportions—these are only some of the most apparent modes of a far more general process of transformation occurring around us. Even in the built environment fundamental "biotechnic" transformations are emerging everywhere. First, processes of unprecedented organic integration of machines and infrastructural systems are taking place at a variety of scales: the integration of sound, text, image, and raw data flows into the single, blended, sensory environment of multimedia; the integration of such microenvironments, or "stations," into permanently activated macro or global systems through digital telephone, cable, or fiber optic networks; and perhaps most important of all, the *phenomenological* integration of the human nervous system into these multilevel, multiscaled networks and environments through the development of so-called "natural" or "fluid" interactive interfaces that use voice, gesture, and touch. These new architectures can be understood only in terms of a new type of concreteness or materiality, one that addresses itself to our organic "modes of attention" and the substratum of human nervous and sensory response.

But this active, informational penetration of matter goes much farther still, first through the progressive saturation of our concrete and mechanical milieus by the microchip, which now augurs to leave nearly every aspect of our inherited material environment at once apparently intact, but also computerized, software-driven, servo-controlled, real-time responsive, and interlinked. Second, as material science delivers its bewildering panoply of active, sensitive, or programmable materials such as shape-memory alloys, glasses, ceramics, and polymers, new hybrid materials with integrated piezo-electric crystals and organic molecules—all of which are capable of changing structure and appearance in response to an array of environmental cues—the very idea of architecture as merely *frozen* music has become at last literally, and not just ideologically, obsolete. The rise of Artificial Life (as opposed to *AI*) think tanks and research institutes, and the remarkable prestige earned by the once purely theoretical speculations of nanotechnology engineers are equally signs of a fundamental shift away from a classical mechanical paradigm.

This new organic, thermodynamic, informational view of architectural form has already begun to emerge today. Witness Peter Eisenman's tendency since 1990 toward the use of drifting, pulsating forms, self-regulating systems, and *continuous*, as opposed to discrete, mappings and transformations; Witness Bernard Tschumi's use of multilevel reaction-diffusion systems (La Villette) in which points, lines and surfaces are transformed, like printed directions on a musical score, into *activities* of intensification, relay and diffusion; And finally, witness Rem Koolhaas's urbanist embrace of fluctuations, instabilities, interference, continuous self-updating systems, and his insight that it is *soft* form, not hard, that bears the maximum of *active* structure.

Yet it is not through architecture alone, not even through the brilliant resolutions of the above examples, that we will come fully to understand our fate and position in the modern technological world; this will occur only in conjunction with a sustained and creative philosophy—understood now not only as a form of knowing but of making as well. Freedom, after all, is not something of which humans can somehow, through operations, be repossessed, but something that must continually be reinvented. If the historical, material, and technical conditions of our being are today inextricably woven together and largely invisible to us, this only means that our conceptual and inventive capacities must be made all the greater. Our predicament can perhaps best be summed up by the following Spanish proverb, once cited by philosopher of technology George Grant:

> Take what you want, said God—take it . . . and pay for it.

1 These are the overriding themes of the work of philosopher George Grant and are directly drawn from him.

Play Time, Jacques Tati, 1967.

PLAY TIME

The city, as Lewis Mumford used to claim, stands among the greatest technological achievements of humankind. The city to which he referred—leaving aside for a moment the difficult question of whether it is the one that persists with us—arguably remains not only our possibly greatest technical achievement, but is also among our greatest cultural and political ones. Nobody would deny the systematic way in which the rise of "the city" deformed human existence, how it multiplied and transformed the brutishness and suffering of a daily life previously unguarded from the indifferent and adverse unfoldings of nature, but now "protected" and subsumed by the new, subtler calamity of unfettered economic rationality. The latter development may in the end need to be seen as inevitable, as it follows from the city's undeniable status as a *technological* object, as a new social assemblage of exquisite and unprecedented complexity, yet one whose engine was never other than the purely and coldly economic. Let us admit, however, once and for all, that analyses of this kind have never told us anything.

The emergence of the city may also be listed among the greatest emancipatory events of the post-Enlightenment era. The sheer social density of the great urban centers—or rather, the almost molecular instability caused by their free and uncontrolled buildup—made possible a thing no less consequential than the political *revolution*, or at least a type of insurrection never seen before: one both "popular" in texture and dynamism and also *driven by philosophical ideas*. This new "alloy," made up of bourgeois intellectual discourse and proletarian mass deployment, could only have been forged from a crucible in which great concentrations of historical substance had come to exist in unprecedented *compression*. That crucible, of course, was the early modern city. Such compression accelerated the dynamo of history not only by multiplying the number of connections and interactions of elements (citizens, institutions, and their mutually engendered affects), but by altering the qualities of their interactions too. If historian and theorist drudges have not ceased reminding us of the uniquely economic aspects of the twin emergence of capitalism and the city, are we not now compelled to develop an alternative culture through which the psycho-erotic, the ludic, and the sovereign dimensions of the new subject, its new social space, and all that is both possible and unsanctioned in these, might be made intelligible?[1] How do we formulate this ever-so-subtly new question?

First published in *ANY* 12, "The Dimensions of Play" (1995).

Following Piece, Vito Acconci, 1969.

I do not expect many readers to buy the claim that what we have to do with here might actually be a *new* question. But the phenomenon of "play," I would argue, *is* a new question; it is always new, just as it is always about delivering up the new. This is why in our work and our reflection we inventors of the vmaterial environment have no other choice but to begin where the work of British psychoanalyst Donald Winnicott concludes.[2] For without any doubt, play is a space with an action and influence of its own, one that opens up only once the small-minded, bureaucratic query of the parent, teacher, or doctor—"Did you make that up or did you find it in the world?"—and has been banished, or more to the point, put *out* of play.[3] Play introduces an altogether different type of productivity (and one not limited to the human cultural world): it allows us a contractual suspension of the gravity of need, frees us of our ingrained meanness, allows us the room to posit hypotheses that will not undergo verification by referees. Only in play do we today find the lost universe of pure truth.

What does all this mean? Perhaps, quite simply, that the product of true play has nothing essential to do with the objects and proclamations that appear to ensue from it, but rather with the invention of the entire alternative world from which those objects and proclamations are drawn. In other words, and there is no reason to shy away from it, play is directly connected to the province of the sacred. Play is about worldmaking *in the absence of verification*. This is not in any way poor science but rather a sustained assault on the status quo of subjugated and managed (and therefore debased) reality. It is the formal, revolutionary destruction of rational or productive *time*. This is why, for Georges Bataille, play was seen as belonging to those catastrophically expensive activities in which a culture participates out of internal economic necessity, such as sex, war, dreaming, festival, etc., while for Walter Benjamin, the activity of gambling (the French *jouer*) may have permitted the industrial worker his only total emancipation from the insidious logic of accumulation and subsumption dictated by workplace machines. Today, more than ever, the spirit and ethic of true play, as a destructive and wasteful activity above all, must reemerge as a necessary ingredient in any encounter of human productive energy with the machinic environment into which it is almost uniformly projected. Sacred space need mean nothing more than the space of sumptuary practice and exchange, the space of "untimely" propositions in the Nietzschean sense, that is, of propositions which go so against the tenor of the times as to appear at once both threatening and almost unverifiable.

ARCHITECTURE AND AIR

Play represents an always new type of clearing; it might be seen as a type of Sabbath in the very deepest sense, the introduction of a lucid, non-productive timeform into a world of crude spatiality, of commerce, and of their relentless technological mastery. Play announces the always inchoate revolutionary project of a world in which events and history—and the cultivation of thick durations—might one day take conceptual precedence over mere places and things. At any rate, play today can honestly exist only as an activity of sabotage, resistance, and wanton neglect of a modern, desanctified world to which it is indifferent *but which is not indifferent to it*.

Many will happily see in these arguments little more than an apologia for the most traditional forms of theological investment in a transcendent state of grace. Yet nothing is further from the case. True play is the cultivation of a radical immanence, the unfolding of cultural, aesthetic, social, and even mystico-erotic values based on no preexisting principle whatever, all this, *and* fully antagonistic to the onward march of economic rationality. Clearly, for play to be real, it must be exceedingly, extravagantly expensive. After all, it is the wagering of existence in the name of a pleasure that is never yet named.

What does the modern world know of such irresistible experience? Very little perhaps, but even that does not mean that we too have to forget. Everyone agrees that our new cities are but vessels of anomie, and even our philosophers are themselves thoroughly dazzled and unable to seize the fugitive element. The emancipations of the modernist city have before our eyes become the nameless gray depredations of the present one. Yet what were—or could now be—these emancipations to us? Did not the city permit the explosive diversification of personality, raise the chance encounter to a sublimated art form, break the stranglehold of church and family on private practice and public morality? Did it not inject a critical dimension of "play" into the social mechanism in the double sense of "free action" and "a sanctified space of pure hypothesis?" Did it not, in fact, *eroticize* the public sphere? The Greeks conceived of a life as an artform, but only in the modern city did it actually become so of necessity. Thus the city, to return to Bataille's exotic thesis, may be seen less as the place where wealth is accumulated and multiplied than the place where energy may be artfully—but profligately and wantonly—*spent*.

In the final analysis, the ecstasy of play has to do with its relationship to beauty. We say that we play music but never—yet!—that we play architecture. Yet the pleasure of architecture (at least of nonclassical types) is never revealed in its image but, like music, in the syncopations of its flow. The theatricality of architecture might actually be understood as an extra dimension *added on* to the merely spatial (instrumental) world, the clearing, so to speak, of the speculative, transient, and fragile realm of play. Just as Greek drama emerged from the one-dimensional line (or half-circle) of the chorus in order to individuate a specific *possible* world from the primordial substrate (the continuity of song), and just as its characters invariably fell back into this substrate in an ecstasy in which, only later, did we agonizingly recognize the ritualization of our own mortality, so now may we see play as an exquisite expression of our assenting to the tragic mode. To play is always to play with death, but at the same time to forge from this anxious proximity a link to beauty which could never otherwise be made. To play (to write a play? to take one in?) is, to borrow a phrase from Shakespeare's time, to "laugh and lie down," to live a life or to give place to form in a transient, even momentary *other* world, and then to die out of it willingly and repeatedly—a world, a form, and a life neither "found" nor "made up," just *lived* in the full immediacy of death. To play is to give shape to what is beyond space, *to time itself*, amid the full and tragic awareness of its waning, to deposit a figure in that which flees, and to know that its deliquescence (and unverifiability) is the lyrical condition of beauty.

1 Certainly by the late nineteenth century, with the rise of philosophical sociology and scientific psychology, certain figures such as Émile Durkheim, Georg Simmel, Jean-Marie Charcot, and others had begun systematically to repertory the whole new metropolitan characterology already exhaustively treated in the novel, in poetry, and to a lesser extent, in painting. In the twentieth century, outside of these "artistic" modes, only Sigmund Freud and Walter Benjamin seem particularly notable.

2 See Christian Hubert's "Playtime," and Joan Ockman's "Architecture in a Mode of Distraction: Eight Takes on Tati's *Playtime*," in *ANY* 12, "The Dimensions of Play" (1995).

3 On the question of legitimation and the ways it is elided by ritual, spontaneous play activity, see Donald Winnicott's *Playing and Reality* (New York: Routledge, 1971). Integral to this problem as well are the socio-historical and narrative modalities of the "carnivalesque" developed in the work of M. M. Bakhtin *(The Dialogical Imagination* and *Rabelais and his World)*.

Despite the impressive formal exuberance and wit displayed throughout its work, the Coop Himmelblau project was born from a darker moment and mood: the ruins of the modernist city at the hands of late capitalism; the privatization of the city and of public equipment generally, and the routinization of design practice, by postwar bureaucratic rationality. As part of a wave of broad resistance to these domesticating forces that included work as diverse as the cinema of Jean-Luc Godard and Michelangelo Antonioni, the literature of Thomas Bernhard and Thomas Pynchon, and, well, Rock 'n Roll, Coop Himmelblau evolved a practice of public provocation that has only recently and reluctantly converged into architectural production.

The Coop Himmelblau project was intended as an almost "pataphysical" assault on bourgeois gravity and matter, a project that strove to address and create nothing short of a new type of "air." Espousing certain ideals—the stealth and volatility of the panther, the lability and unpredictability of combustible matter and flames, and above all, the paradoxical colossalness and buoyancy of clouds—they upended many of the clichés of western culture's romance of how physics is connected to economic and social reality. Gravity is shown to be a myth whose purpose is to naturalize "work" in the body of the worker (mass x displacement), to envelop the human in a grid of straight lines (the alibi of the parallelism of two falling bodies—in fact these are never parallel but always converge at the center of the planetary sphere), and to shape human subjectivity by anchoring it to an internal "center of gravity" that is made to appear as a Newtonian universal.

The cloud becomes the very image of human emancipation from this Newtonian bondage. While their interest in the shifting imprecision of these mesmerizing and continually varying natural forms serves as a main engine in the work, it carries well beyond this too. Clouds were the way that Karl Popper famously referred to a different and necessary style of scientific knowledge (he derived it from the analysis of gases), appropriate to a world in which unpredictability and therefore openness reigns. The work of Coop Himmelblau celebrates the joint human and scientific emancipation from the "clockworks" of classical physics by seemingly denying the laws of bourgeois physics themselves.

Seagram Building—Ludwig Mies van der Rohe, Hiroshi Sugimoto, 1997.

RADICAL ANAMNESIS
(Mourning the Future)

It is customary today, and one hears this even from thinking people and not only academics, to dismiss points of view and systems of thought as "nostalgic" whenever they attempt to summon the nobility of past formations or achievements to bear witness against the mediocrities of the present. To use the grandeur of a selectively remembered past to carry us beyond, and to help us see through, the *nombrilism* of the present may be the sole value of history, and memory would be nothing other than the technique by which this reservoir of creative potential is tapped. I, for one, firmly believe in the power of memory—in *radical anamnesis*—as a genuine act of political resistance in our structurally amnesiac and personality-dissociating social world.

Once, while writing about the capacity of organized matter to demonstrate rudimentary features of life, I associated agency—the capacity to overcome inertness and torpor and to *act*—with intelligence, and proposed that intelligence was originally made possible by memory.[1] Drawing on one formulation of this idea, German scholar Anselm Haverkamp recently opened a special issue of a design magazine dedicated to the problem of memory.[2] I was especially moved in that issue by Haverkamp's evocation of mourning as an indissociable aspect of the action of memory. If memory is creation and invention—creating links from one place and thing to another—rather than mere reproduction, as more conventional thinking would have it, then it is all the more startling to link it to mourning, the affective and existential connection of a subject to loss and deprival. History, as we know, both gives and takes away. We know what it gives us because for the most part we are left to wallow in its endowment, but how do we know what we have lost? How can we construct an image of what is possible for us (as distinct from what is already actual) except by a mental and existential act of projection outside of *this* present?

Thomas Pynchon, in his 1973 epic *Gravity's Rainbow* (a work whose evocation remains a defiant, repeated act of "nostalgia"—or anamnesis—on my part), invokes the concept of "temporal bandwidth," a literal scale on which to measure both the onslaught of twentieth-century modernization and the progressive disintegration of the main protagonist's personality. Temporal bandwidth measures the degree to which one's being remains psychically and existentially linked to its past and oriented toward its future. The more cut off from these it is (as the temporal bandwidth shrinks), the more its being becomes concentrated in an increasingly detached present, the more unmoored it becomes and the less it is able to rely on stabilizing continuities of coherent unfolding or becoming. This disconnected state corresponds to

First published in **ANY** 15, "Memory, Inc." (1996), guest edited by Anselm Haverkamp.

the "schizophrenic" model according to which everything is disarticulated from everything else—with neither pattern, causality, nor whole. At the opposite pole stands the "paranoid" model according to which every thing and moment is inextricably and seamlessly *connected* to every other. Both states represent fatal rigidities into which we must, as historical beings, avoid falling. The antidote is the flexibility afforded by controlled remembering, not only of what we were but, in the same emancipatory act, of what we might be.

Mourning may be among the most underrated of philosophical postures. To mourn is to lay down a path to nobility and to creation intolerable to, and free from—for that moment at least—the tyranny of any actual present. To mourn is not to refuse loss, but rather to keep its contents (non-neurotically) active in a present which is ontologically hostile to it. In Haverkamp's brief formulation there lies one clear example of what I call radical anamnesis: the imagination's escape from the sterile logic of what is.

The architect in the twentieth century not only builds worlds but must confront the world that is already built, as well as the subjects and the mores that are forcibly built within and by it. Some attention in architectural circles is currently shifting away from the brute artifacts of building toward the predatory or progressive routines that they unleash in the social sphere. This new interest in the multiple forces that combine to form subjects, that condition our capacities to live and to call ourselves citizens in a world of increasingly cruel and encompassing sets of logics (media, economics, technology), is a realm that must today be occupied by both thought and practice. It has always seemed to me strange that architecture—that is, architects—with few exceptions, fear the big picture. The parochialism that perennially threatens the discipline, the timorous assumptions of editors, and the grandstanding and false bravado of successful practitioners all feed into the fear and into the narrow ambitions of architects and design intellectuals as cultural producers, social and political actors, and shapers of history. It also testifies inexplicably to how short architectural culture in this country, despite the Herculean efforts of a Lewis Mumford, a Jane Jacobs, a J. B. Jackson and a Reyner Banham, still falls. On the other hand, memory as a deliberate act of will projects one squarely into the stream of history, beyond the shackles, the smug certainties, and the petty stakes of the present. Through (selective) memory the future becomes possible, a future that the past could not think and that the present—alone—dares not.

1 See "The Cruelty of Numbers" (this volume, p. 92).
2 *ANY* 15, "Memory, Inc." (1996), guest edited by Anselm Haverkamp.

WHO'S AFRAID OF FORMALISM?

Tea Ceremony. Submitting freeformality to the ecstasy of rules.

True formalism, we imagine, has been under siege for nearly as long as it has occupied—and mostly merited—the forefront of rigorous analysis in the arts and the inexact sciences. But in truth, this has not been the case. Only the poor or degenerate formalisms (the merely "formalistic") have drawn cogent challenge from so-called higher-minded, more ecumenical modes of analysis. "Poor formalisms," I would claim, are really just *unextended* formalisms: parodic analytical methods derived from the great and genuine aesthetic and epistemological innovations of modern, avant-gardist tradition, but which have simply forgotten that that is what they are. For the poverty of what is today collectively referred to by the misnomer "formalism" is more than anything else the result of a sloppy conflation of the notion of "form" with that of "object." The form problem, from the time of the pre-Socratics to the late twentieth century is, in fact, an almost unbroken concern with the mechanisms of *formation*, the processes by which discernable patterns come to dissociate themselves from a less finely-ordered field. Form, when seen from this perspective, is ordering *action*, a logic deployed, while the object is merely the latter's sectional image, a manifest variation on an always somewhat distant theme. The form of the object (or the form of the expression) and the form of the theme (form of the content) are in truth in continual dynamic resonance, and, when grasped together by formalist analysis, open up onto a field of limitless communication and transmission.

What I call true formalism refers to any method that diagrams the proliferation of fundamental resonances and demonstrates how these accumulate into figures of order and shape. The very idea that the figure of a façade, the plan of a villa, or the marquetry of a given urban tissue might enfold within it a resonant, transmissible logic of internal control, one that can be at once dissociated from its material substrate *and* maintained in communicative tension with it, was once an assertion of great contentiousness. The moment of its rigorous demonstration became one of the watersheds, not only of modern aesthetics, but of modern science and philosophy as well. Colin Rowe's work in architectural analysis laid one of the cornerstones for the possibility of the emergence of a true formalism.

Yet the configuration of the contemporary polemic is misleading. It holds, among other things, that an enlightened science of "meaning," of ideology, or of commitment can and ought to be mustered as a palliative and corrective to a sterile, abstract academicism that seizes only the visible but lifeless schemata of things, one that weaves its pallid array of skeletal elements into a fraudulently brilliant, self-fulfilling, but world-denying view. Such a position might be partly valid—it could be seen as attacking the poor formalisms of the object—that is, if the latter were not in fact but the concoction of the former's own flawed understanding. Indeed there is not, and never has been, any such thing as "meaning" or "ideology," not, in any case, one separate from the physics of history and power: a physics, not incidentally, which is always a physics of *forms*, be it the form of an idea, the form of an epoch, or the form of a tool. True formalism holds out for us the real possibility both for a pragmatic description of historical emergence (why this object, institution, or configuration here, in this place, at this time, and not that?) and the superseding of the tired and wooly metaphysics that continues to dog thought today, the metaphysics of the signifier.

Formalism demonstrates first and foremost that form is resonance and expression of embedded forces. The best local formalisms—those of Rowe, Henri Focillon, Arnold Schönberg, Michel Foucault, etc.—show that these embedded forces are themselves organized and have a pre-concrete, logical form of their own. The dynamic relation between these two levels of form is the space where indeterminacy or historical becoming unfolds. Extended or true formalisms are different only in that they also describe relations of resonance and expression *between* local forms or form systems. This is why most anti-formalists are essentially but poor formalists themselves; they see only the shell of object-forms and sad enclaves of inert matter, never the resonating *field* of wild, directed formation. The great formalists, on the other hand, have always been able to peer *into* the object toward its rules of formation and to see these two strata *together* as a mobile, open, and oscillating system subject to a greater or lesser number of external pressures. The manifest form—that which appears—is the result of a computational interaction between internal rules and external (morphogenetic) pressures that, themselves, originate in other adjacent forms (ecology). The (pre-concrete) internal rules comprise, *in their activity*, an embedded form, what is today clearly understood and described by the term *algorithm*. Algorithmic formalism (the most dynamic, extendable kind) was an invention of Goethe's and remains the basis of all robust, generative formalisms (including those being used today in computational biology).

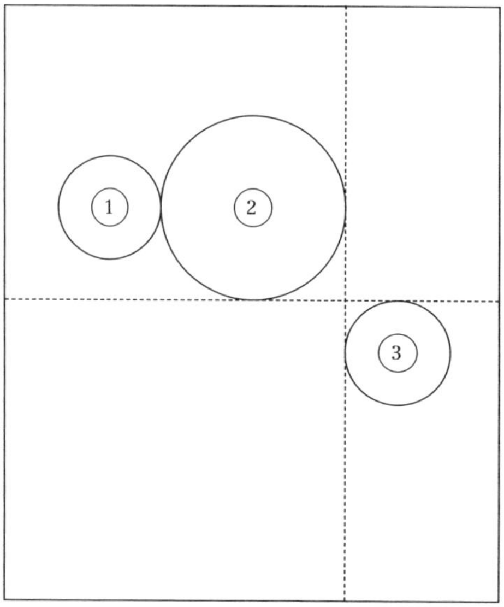
1
2
3

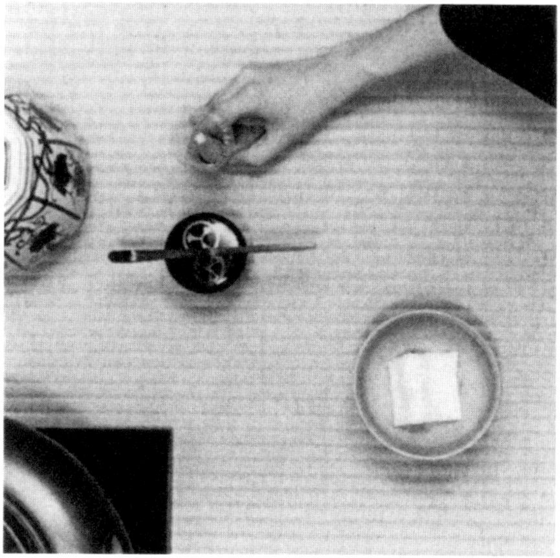

Among other things, Goethe posited the concept of a "type" as an abstract formative principle to be acted upon by other primary transformative processes. This may well be the source of a misunderstanding today regarding the role of generative or "deep structural" elements in designed systems and design processes. For the type concept is *never* a development of a supersensuous Platonic *eidos* (one intuits in such contemporary claims the tendentious, reductive influence of Derrida), but rather is related to a dynamic inner intelligibility (the *eidos* of the Physicians, linked to *dynamis* or power), or to *actualization* as in the formal causes of Aristotle (*eidos* in its relation to the entelechies or to *energeia*). Type, in Goethe as in Rowe, is at least partly active, and it is on this active aspect that we need to concentrate if we wish to give place to new extended formalisms.

Ernst Cassirer once said of Goethe that his work completed the transition from the *generic* view to the *genetic* view of organic nature. He was referring to the break from the tabular space of the *genera* of the Linnaean classifications and their emphasis on what is constant and fixed to a *generative* space where the processes of coming-to-be are given shape. Goethe's formalism, like all rigorous and interesting ones, actually marks a turning away from the simple structure of end-products and toward the active, ever-changing processes that bring them into being. With any luck, twenty years from now, one will be able to make the same claim for Rowe's work in architecture that Cassirer made for Goethe's in science. And should in fact this not come to be, it will be far more the fault of the one-dimensional semioticians and ideologists who propagate the cliché of the "social construction of meaning" than of second-rate poor formalists who merely trivialize a powerful method and inadvertently lend credence to the airless arguments of the former group.

WHAT'S EATING CHARLES GWATHMEY

(and his Generation)?

If the chorus of testimonials assembled in the July/August 1995 issue of ANY magazine[1] may be taken as a fair and representative probe—as I believe it to be—then one might well conclude that the fate of American formalism finds one damaging exponent in the work and career trajectory of the Gwathmey Siegel & Associates office. Perhaps. But of that entire generation, in fact, whose youthful rigor and brilliance were first celebrated and publicized in the (disingenuously) modestly titled work *Five Architects*, it is at this point well and fair to ask publicly what, for over a decade, has not ceased to be murmured in intimate professional circles: "What happened?"

What, that is, happened to the noble *historical* project, the committed break with traditional pieties, the transvaluation of values, the spirit of risk, research and redemption of, above all, a new human "type?" Formalism, after all, may well have been the first real revolution in architecture: the radical transformation of public forms would disrupt the silent progress of reactionary social programs; the manipulation of plastic forms would call forth, with a whole new menu of sensual cues, a new and different human animal whose eros and freedom could be left to pure invention, variation, and play. Alas, this description sounds like no more than what it is: an incantation of the all-too-remote "aristocratic" decades of the twentieth century, the 1920s and 1960s. Yet why are such historical episodes so easily excused or dismissed? If this were only weary and feckless Europe, or insular, late-awakening Asia, one could understand. But not in North America, *the* modern continent, the very synonym for modernization and the embodiment of the twentieth century itself. Why did many of our formalisms fail (actually, they didn't fail, they degenerated), pitilessly taking Charles Gwathmey's down with them?

One reason clearly was the weirdly exaggerated importance given by the American modernists to only limited aspects of the work and ideas of Colin Rowe and Rudolf Wittkower. The result was a directionless formal hyperbole left unchecked and unmodified by a purview too intellectually parochial to comprehend and assimilate the profound and especially related developments in adjacent fields like art, literature, music, and linguistics. But this certainly goes only a short way towards explaining the "toys, teapots 'n 'toons" aesthetic of later Michael Graves, or the inflated phantasmatic *bibelots* of John Hejduk, or even entirely the aimless and often soulless "modernistic" tinkering of Gwathmey's later work. At least Richard Meier, for a long time, remained driven by a passion for the paintings and ideas of Frank Stella, and, if he and Stella went down, well, at least they went down together, swathed in remarkable but ultimately cloistral mannerisms of their own making. And then there is Peter Eisenman, tireless irritant and provocateur, who, for better or worse, has probably been unsinkable only because he remains committed to novelty, is unafraid of failure, and because he still recklessly drives his vehicle with all the rearview mirrors removed. (The book *Five Architects* may well have been a *cadeau empoisoné* from Eisenman to his four, more ambitious but less robust friends, a set up, a twenty-year practical joke.) Among Americans of that generation the only truly impressive survivor of DAMS (Degenerative American Modernism Syndrome) is probably Frank Gehry, a decidedly "lite" practitioner in relation to the urbane Five, but with a modernist pedigree of early built work as refined, graceful, and intelligent as anyone in the practice. The youthfulness and exuberance of Gehry's late work, like Eisenman's, possesses more experimental vigor than that of architects half his age, if also (like Eisenman's) a perfectly shocking number of bellyflops. But the project and spirit are there.

So what, specifically, is eating Charles Gwathmey? Well, it's not that he doesn't give a damn what the intellectuals think, for even these days his lectures remain self-consciously saturated with the once highfalutin' language of academic 1960s formalism. (The effect is all the more strange when one considers the painfully tarted-up recent work to which the threadbare rhetoric is being applied.) The most significant impression one is left with is that his imagination has not accumulated an ounce of new baggage, complexity, dimension, or ideas in twenty-five years.

To be a modernist is to be something more than a great compositionalist, a virtuoso of the modernist plastic lexicon. Gwathmey is undeniably a gifted architect who played in the league where one's work makes of one an historical figure. But history stingily tries to take back what, at a later moment, it fears it may too hastily have paid out. To be a modernist is to be *existentially* committed to history (not just professionally), to believe against all contradictory evidence that one's work and one's existence are potential agents of historical transformation. The idea that something historical can be had cheap,

First published in ANY 11, "Legitimate Transcriptions: The Early Work of Charles Gwathmey" (1995).

or for free, is sadly something that has of late become part of American tradition in an implicit but corrosive way. For in architecture there is another background figure—a bit like Rowe and Wittkower—who enjoyed a substantial, if incomprehensible, influence on Gwathmey's generation and whose influence is largely responsible for this diffuse but unmistakably "grifter's" approach to history and design. That figure of course is Philip Johnson, gadfly and interloper in the American architectural world, who, with a thin endowment of talent and intelligence (and moral substance) managed impressively to make up for these deficits with family money, an instinct for manipulation and intrigue, and an inexhaustible groupie mindset. While his fortune and social connections were sufficient to secure an "exclusive" option to import and broker the Modern Movement to the American public, his intellectual depth was certainly never adequate properly to do so. The result is well-known: a dilettante's version of an historical process, a debasing reduction of cultural struggle to a privileged boy's parlor game, at best, a dapper new style. In the postwar exuberance in America, history was still traumatically difficult to digest, but style was as ravenously consumed as during the depression years. Johnson, already a willing parody of a one-dimensional de Tocquevillian character, happily presided for decades over this world, perpetuating just this climate until it became a veritable American tradition.

This sad and unsavory beginning certainly changed the American intellectual and design landscape in ways that can only be lamented today. The homeless, seemingly unrenewable, and, if true, unredeemable American formalism for which Gwathmey must share the blame unquestionably belongs, at least in part, to a tradition made possible and fostered by Johnson, by his limited purview of the world, and by the disproportionate effect that this has had on the way much of America has consumed architectural culture since World War II.

What's eating Charles Gwathmey is probably not so different from what was eating Gilbert Grape: the presence of an inflated human behemoth so entrenched in (a piece of) architecture that it has virtually become identified with it, and an unseemly obstacle to the free development of all who are related to it.

Yes, something is eating Charles Gwathmey and his generation, but we are all at risk of becoming victims to it.

1 *ANY* 11, "Legitimate Transcriptions: The Early Work of Charles Gwathmey" (1995).

The Endemic Identity Crisis.

Of all the shibboleths by which we are tested today, none is so relentless, nor so deliberately menacing to our complacency, as the slogan *globalization*. It is no surprise that, like Saint Augustine, we are fairly confident that we know what it is, until someone asks us to describe it. We then start to blather—to sound like economists, international relations pundits, political and social theorists: in other words, we get lost in a tangle of boilerplate, vivid anecdotes, and hollow abstractions. As Thomas Friedman argued once in a *New York Times* editorial,[1] the term globalization names nothing more than a napkin game that allows us to define and declare where we stand—vis-à-vis free trade, integration, separatism/protectionism, laissez-faire capitalism, social security, etc.—in relation to the positions of others and according to the most deceptively familiar criteria. It does little, however, to confront a set of novel and often brutal processes already unleashed, entirely irreversible, and for the most part beyond current grasp and control.

Nor does globalization mania appear to be making us more clever. Clearly one of its effects is to have begun to destroy, in our minds and to a lesser extent in our world, the classical distinction between *market* and what used to be called *civil society*.

Economic rationality is worryingly becoming the unchallenged lingua franca of the New World Order—hip and hardnosed and so rife with exclusionary protocols that it has become downright sexy, even among left-wing thinkers. Among the many causes contributing to this state of affairs is the passing of the Great Ideologies from public view, the demise of totalitarian and communist regimes in their full world-national splendor and explicit representation. Yet only an economist's habits of literalism and "misplaced concreteness"[2] (to quote philosopher Alfred North Whitehead's attack on the atomist's science) could imagine that the *manifest* disappearance of a highly entrained ecological object or phenomenon from the historical landscape meant that it had *actually* gone away.

In the dense matrix that is our socioeconomic world, things do not simply and suddenly disappear but rather *change mode*—they spore. The terms *dormancy* and *latency* were formerly used to give image to these phenomena, but even these descriptions suffered from the mechanistic fallacy through their tacit assumption of the stasis and passivity of the agent that had been transformed.

INFRARED, THE TREASON OF LANGUAGE, AND THE FAILURE OF THE GEOMETRIC IMAGINATION

To make an object "vanish" from the socioeconomic manifold is not to erase it but rather to displace it to another level or scale of action. The portion of the dike from which its concrete presence has been withdrawn then becomes unstable, presenting a flood point to forces that have long awaited their moment to emerge. (Do economists really wish us to believe that the "new" forces are created *ex nihilo*, on demand, and instantaneously, the instant a "market" for them emerges?) The realm from which the so-called new forces come is no other than the one to which "old" ones return: the realm of the "infrared" and the "ultraviolet."

If we insist on crude appearance as a criterion of understanding, we will miss all that has to do with remote agency, because in social space remoteness denotes not a distance of measure but a distance of modes. What we need today is a clear set of techniques with which to create images of the action of forces, *to make the invisible enter into intuition and action*, to grasp the concreteness of what around us is not represented or even representable. Gestures against economic reductionism are today everywhere afoot. Among the most controversial are Samuel Huntington's *The Clash of Civilizations and the Remaking of the World Order*, a work that argues that cultural structures, rather than economics or ideology, will be the sources of future political conflict, and George Soros's February 1997 *Atlantic Monthly* article, "The Capitalist Threat,"which argues that free markets not only cannot represent but, in the absence of counterforces, actually undermine open and democratic social structures.[3] Huntington's work reminds us of the hubris we commit when we fail to remember that modernization and economic liberalization are never happy synonyms, but rather wishful, simplistic thinking about a complex and viscous world that has no interest in, nor ability, to renounce local conditions or persisting archaic endowments; Soros reminds us that even economics is determined by reflexive structures and feedback loops that insidiously shape the very world it wishes only to represent. Globalization mania is itself a significant current threat to clear and nuanced thinking.

Among the most extreme real examples of raw development in an unbridled market environment is the pattern of aided growth that characterizes China's Special Economic Zones. While renowned economists such as Paul Krugman and others debate whether Asian growth is real or illusory,

First published in ANY 19–20, "The Virtual House" (1997), a special issue on the virtual dimension and its significance in philosophy and design culture. The "infrared" serves here as an elaboration of the theme rather than an analysis of it. The analysis of the virtual was a principle aim of an earlier work, my *Architectures of Time* (Cambridge: MIT Press, 2000).

RATIONAL IRRATIONALISM

Shanghai Stock Exchange.

During the 1970s, when it had become an obligatory affectation of macho bravura in architecture to manipulate the ruler and straightedge with something akin to a virtuoso performance—note Libeskind, Tschumi, Eisenman and others—Zaha Hadid eschewed the posturing of these anti-classicizing classicists by inventing a new kind of line entirely. This new line was at once the centrifugal expression of a pulse of energy fleeing from a moving point—as in a universe not of grids but of vortexes—but also the reinvention of the line in relation to a new type of eye. On the one hand, Hadid's softly arcing lines that compress such enormous energy into their subtly inflected bends are generated by the hand and body itself, by the inescapable rotational dynamics of the hips, shoulder, elbow, and wrist as they coordinate to extend the pen through space yet also leave a powerful trace of their orbital roots. Far from seeking to hide these organic geometrical foundations, Hadid extended and exaggerated their seductive and radical qualities, to the infinite perplexity and fascination of her self-proclaimed avant-gardist colleagues. Those most threatened by the unexpected (and untimely) affirmation, within Modernism, of the body and its lines of flight, dismissed this type of work as "paper architecture." But the question of the eye was a critical weak point of the modern and Modernist tradition, and this is where Hadid's ace resided. Modern space was born of a rational connection to optics, the moment when Brunelleschi and others formalized the modern perspectival approach to drawing based on a vanishing point and the straight lines that emanated from it. But the universe was not like this at all and every Renaissance theorist knew it. The cosmos was rather a play of orbits and attractions, bends and perpetual influences, a continuum of ongoing angular interaction and modification. What's more, the eye itself was dual and mobile, its imaging surface spherical, and the ground to which it is anchored was defined by a horizon and an azimuth, both of which bend and bend visibly and meaningfully. It was in many ways Zaha Hadid who restored this tension and reality to the world of space-making.

Hadid's vertiginous work both transposes and displaces the very horizon that serves as our orientation point in the world. Her curves arc ever-so-slowly as they careen across the canvas or page as if to mock the straight lines that they partly portray, but also to free our intuition from the many regimens of conventional orthogonality to which our modernity has subjected us. Until recently, perhaps the last seven or eight years, her lines never ventured to risk more than a single inflection; today, one finds two- and even three-part arabesques as the work becomes ever more plastic, and the development of the semi-free "line" becomes ever more extended into developing the possibilities of sheets and surfaces and even three- and four-dimensional flows of space and material.

certain brute facts undeniably remain. These too can and must be accounted for, if not with the statistical quasi-proofs of the economists then in the concrete terms of historical analysis, urbanism, sociology, and aesthetics. Economists themselves admit to not knowing anything for certain until long after it has been proven and, for that very reason, becomes no longer of practical interest. To grasp the concrete realities at the moment of their emergence, while they are still malleable and receptive to intervention, requires a different, albeit defective type of "science." An interesting example of such an experiment is the one being run at Harvard's Graduate School of Design (GSD) by architect Rem Koolhaas.[4] In 1996 Koolhaas set up a workshop to investigate, by means of a relentless sampling and accumulation of concrete detail, the empirical "facts" that comprise the Pearl River Delta economic initiatives. Among the ideas of enormous explanatory power introduced in that workshop was that of the "infrared" developed by Hungarian researcher Mihai Craciun. If "red" is understood as the visible face that international communism traditionally wished to present of itself, it chose a deliberately provocative and noticeable section of the visible light spectrum with which to puncture social and psychological space and in which to embed values such as justice, utopianism, and revolutionary change. Red, of course, also came to mean the entire apparatus of communist power and administration and the secondary effects of these as they became more and more embedded in material and moral life. It later became a conscious tenet of the Party that to continue to govern effectively would require the concealment of its ideology—its "disappearance!"—or, more accurately, *its passing into the nonvisible regions of the spectrum*. Craciun's infrared represents the strategy of passage to an invisible (i.e., ideology-free) or embedded communism. Subsequently the Chinese Communist Party fathers, under Deng Xiaoping's reforms, would unproblematically embrace "market competition" as a pragmatic update of Marxist-Leninist "class struggle." The Special Economic Zones and the rhetoric that necessarily accompanies them are in every sense a product of this fundamental dynamic. The communism—the red—doesn't disappear; it is merely displaced and modally transformed. It has become "infra" red.

During a review of the Harvard group's research an enormous sum of data was presented to the jurors over a nine-hour period (much of it describing a reality that few of us would have believed had we not been presented with its avalanche of evidence). Among the standard questions asked by several generations of economic historians are, "Why did capitalism emerge in the 16th century, and why in Europe rather than elsewhere?" In other words, what specific set of necessary conditions were in place that made those developments possible in that place at that time, even though they were not operating explicitly in a well-lit foreground? The answer, to this day not definitively supplied, directly interrogates the problem of remote agency. It has to do with shapes embedded in the "infra" regions of the historical spectrum.

During the ensuing discussion at the GSD I had occasion to propose an hypothetical model to account for the strange set of aberrations that characterize urban developments of the Pearl River Delta economic zones. I suggested that development in late twentieth-century China was merely following lines of "potential" established by its very different historical trajectory, specifically by the structures that had been embedded in its "genetic" space through historical deposits made during its Taoist, Confucian, and Communist past. Once a (universalist) capitalist seed was planted, it would grow, but only along *indigenous* pathways accorded by the geometrical data available in its virtual storage banks. My hypothesis, however, was not meant to explain the *limits* of Chinese capital, but rather the delirious and apparently almost frictionless blooms or flares of development. The theory sought particularly to explain the bizarre yet seemingly plastic relation between micro- and macroscopic growth tendrils, the relation between rigid molar patterns and chaotic molecular ones. I hypothesized that over time a variety of contradictory blueprints had been embedded in genetic or infra space in China, but, though invisible, had remained active (potentially) and "cue-able" nonetheless. The infrared, for example, permitted the massive centralized bureaucratic technologies and apparatuses of the planned state to remain available (those very things that the market-obsessed West has never had as part of its endowment), to be called upon when necessary or when developmental routes became blocked along other, more traditionally "liberal" capitalist pathways. (This also accounts for the specific "syndicalist" mafia style of contemporary Russian capitalism, which might be said to be drawing on Tsarist conventions and structures as well as Party communist ones.) The robust and frictionless Chinese growth patterns of the mid-1990s are unimaginable in the West, precisely because our historical-genetic endowment—our "infra-space"—is different.

In the Chinese example, I surmised, the “red” had been effectively transmuted into an engine of enormous torque and displacement, providing from its invisible perch a macroscopic top-down framework for capitalist infill to which no bottom-up or “orthodox” capital has ever had access. Confucian and Taoist modalities naturally contributed other morphologies to the mix (Singaporean development and authoritarianism are readymade examples of the first), but it is unnecessary to speculate on these here.

Among the jurors at Harvard that day was the Marxist critic Fredric Jameson. Although Jameson revealed few opinions on the China data, he took clear exception to my analysis and cited the early work of Gilles Deleuze and Félix Guattari as a better example of where one might begin to achieve some understanding of the bizarre PRD phenomenon. But Jameson followed up with only a murky reference to a “flow model” developed in Deleuze and Guattari’s *Anti-Oedipus*.[5] His comment struck me as a somewhat desultory attempt to recuperate an economic model long ago and egregiously surpassed not only by world developments but by the considerable refinements and elaborations of the authors themselves (e.g., in their second volume of *Capitalism and Schizophrenia, A Thousand Plateaus*.) Not to recognize either in the Pearl River Delta archive or in the work of Deleuze and Guattari the resolute development of polyphonic, imbricated “geological” structures, of controlled random movement and combinatorics, of unpredictable approximations as a generator of simple evolutionary behavior, in favor of a crude (however clear) base-superstructure model and its last-gasp, post-May ’68 flow update, is a considerable failing in both intellectual style and thought.

The electromagnetic infrared may not be perceptible to the eye, but it is often discernable by the skin and can be precisely tracked by contrivable instruments. The cultural-political infrared is discernable (in the world and through the senses) as a set of pressures that, once incarnated, become visible through vector analysis, as in physics when one works back from the curling tracks in a bubble-chamber diagram to reconstitute the forces once enveloped in the smashed atom. In other words, when surface appearances are treated as a set of resultants from deep in infra-space. We must not be led astray by the languages of theory and criticism, by the face-value evidence of mere appearances, indeed by anything that reassures simply by its conceptual or practical stability. *The infrared is a shape* that causes and inflects other shapes. Its presence is that of a morphological seed growing holographically within, and leaving its characteristic distorting signature on the shapes around us, within the world of concrete appearances. Merely to read these appearances off the official roster sheets of historical reality and to interpret them with the very concepts that have been served up from within the same worldview is to be caught up in the treacherous tautologies of language and representation. In geometry, however, there is hope, but only if "reading" can be elevated (sublimated?) to a process of total intuition, to a defective science of sampling concrete facts, an inventory not of mere things but rather of the distortions that have been laid into things, of the endless play of influence and agency. The geometric imagination must be cultivated in the modern world, and the discursive one relatively discouraged. Our theorists are serving themselves, and their products are leading us into blindness.

1 Reporting from the 1998 Davos World Economic Forum. Friedman has since published a surprisingly simplistic book-length work on globalization, *The Lexus and the Olive Tree* (New York: Farrar, Straus, Giroux, 1999).

2 See also "The 'Avant-Garde' in America, or The Fallacy of Misplaced Concreteness" (this volume, p. 74).

3 Huntington, *The Clash of Civilizations and the Remaking of the World Order* (New York: Simon and Schuster, 1997). Huntington is chairman of Harvard's Institute for Strategic Planning. See also George Soros's subsequent book, *The Crisis of Global Capitalism: Open Society Endangered* (New York: Perseus Books, 1998).

4 The Harvard Project on the City. See *Great Leap Forward* (Cologne: Taschen, 2002) and *The Harvard Design School Guide to Shopping* (Cologne: Taschen, 2002).

5 Gilles Deleuze and Félix Guattari, *Anti-Oedipus: Capitalism and Schizophrenia* (Minneapolis: University of Minnesota Press, 1983).

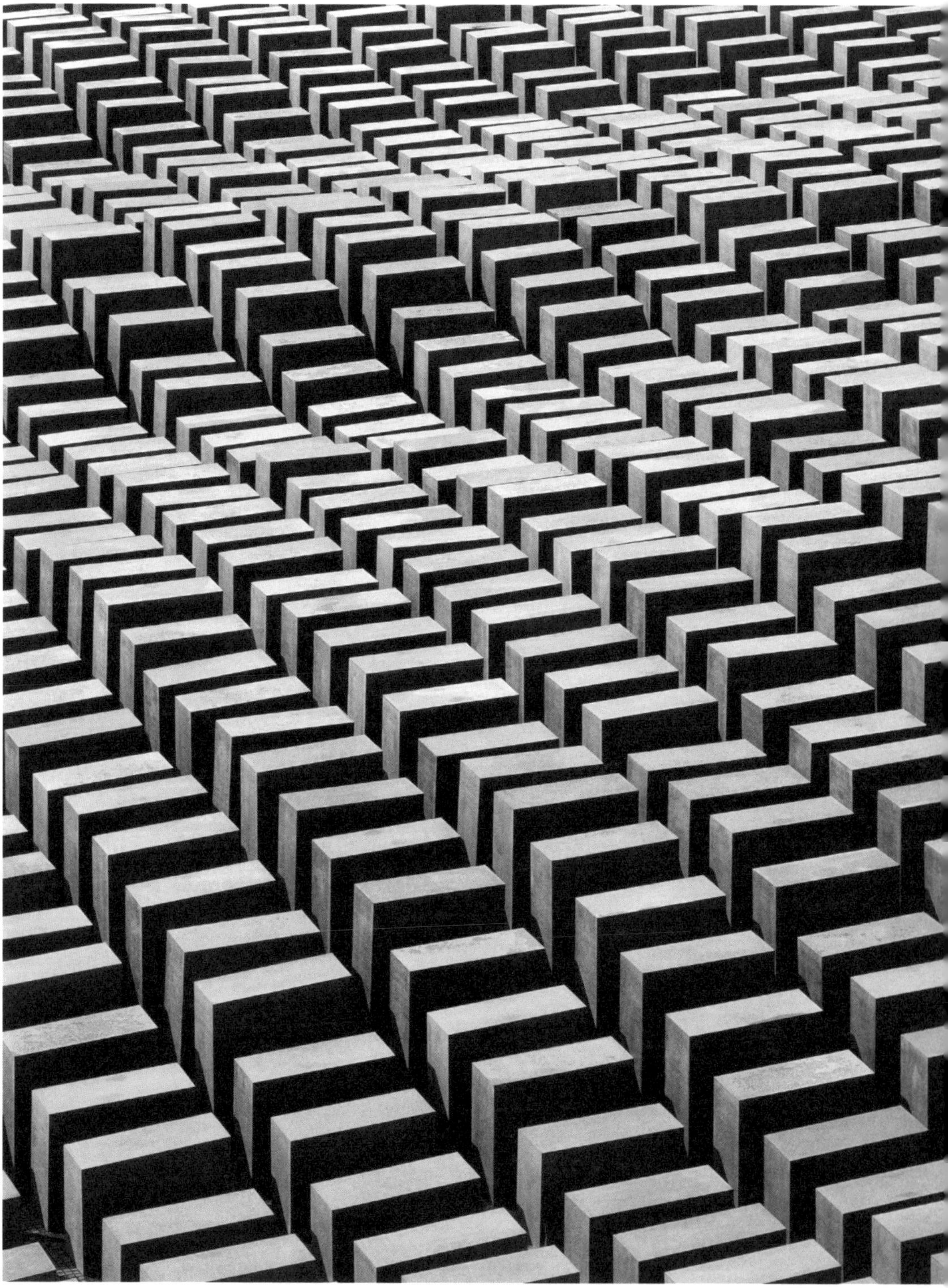

THE GEOMETRY OF CINDERS

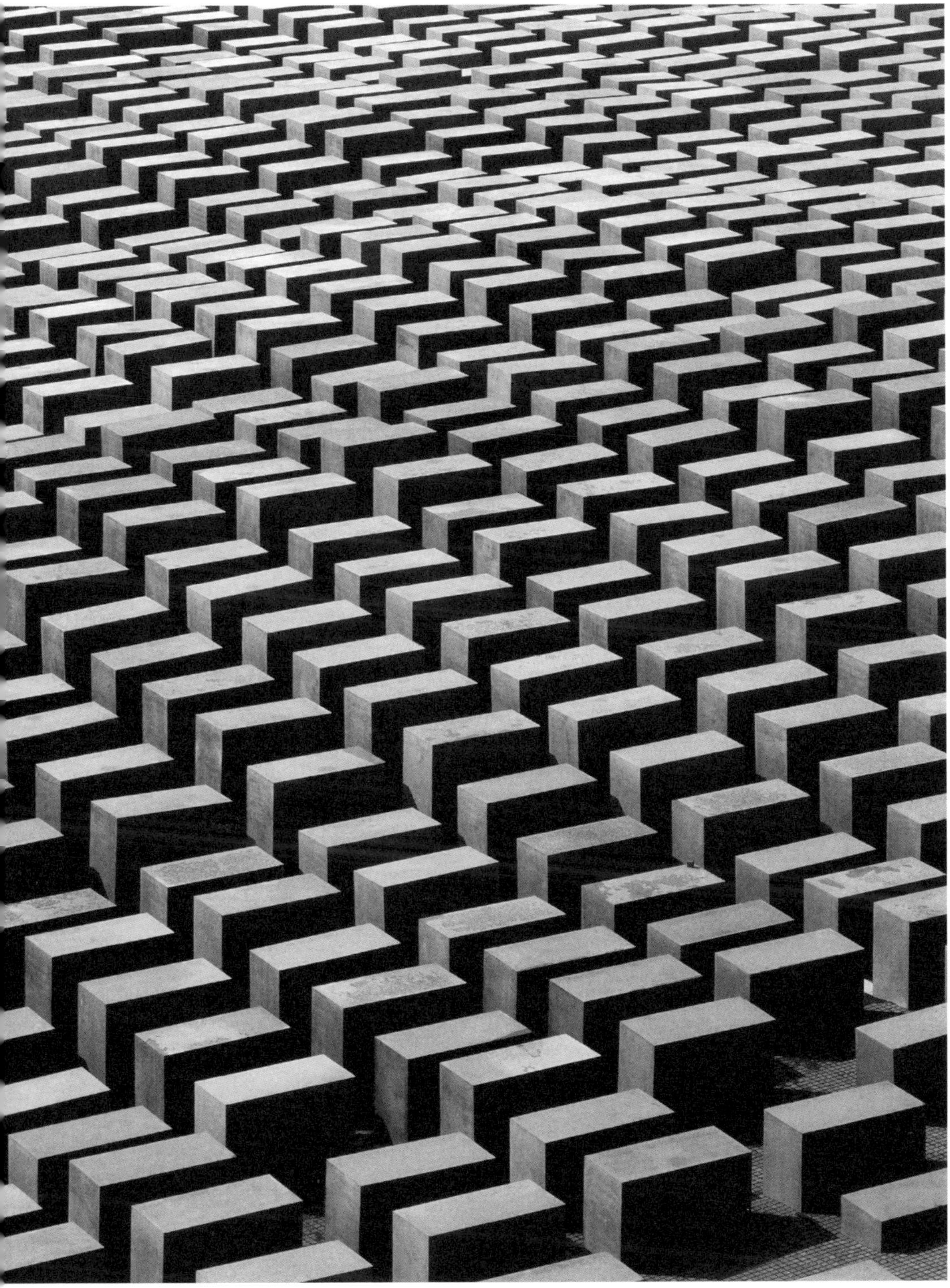

Beyond the expanded field: Memorial to the Murdered Jews of Europe, Peter Eisenman, 1997–2005.

When built, Peter Eisenman's proposal for the Memorial to the Murdered Jews of Europe in Berlin, Germany's Holocaust memorial, will take its place among the largest artificial monuments in the world.[1] Its almost brutal reticence functions in stark counterpoint to its colossalness, amplifying both the silence and the pregnance of its overall form, and laying the foundation for the work's general infrastructure of tensions. Like a massive, gridded shroud laid deliberately over a lifeless landscape, the global fabric of the project's top layer nonetheless appears to waft gently and gracefully over the earth, following and dignifiedly transmitting the muted undulations below as if this were its single and quiet purpose. All successful monuments that address serious themes have in common this particular approach: they *evacuate* information from a landscape rather than introduce more into it, they project a disjunctive synapse into the urban-historical continuum, a kind of *interruptus* that makes way, however modestly, for the action upon us of the sublime. For better and for worse, this always has the effect of desecularizing a space, of removing it from the banal world of immediate economic activity, and delivering it to a different economy, to that of geological, biological, or astronomical time. Only here does the human animal forcibly rediscover itself against a backdrop that is larger and slower than itself, only here does the human experience itself first and foremost as an existence, set off from the swathe of continuous petty ministrations that makes up modern life. Herein is one of the great forms of relief afforded by geometry: it has the capacity to deliver us to a pure and multiple time.

All non-statuary monuments are arguably *memorials* insofar as they free us from routinization and from the million insidious bonds to the quotidian that normally sap and tyrannize modern attention. This was always the power of aesthetic abstraction, and it is why the modern world would have had to invent it even were abstraction not already so intimately a part of its logistical and economic project. Abstraction then lies on both sides of the historical-political continuum: it is at once State Apparatus *and* War Machine (to rally the contemporary jargon), oppressor *and* resistance. History and geometry, despite the narrow worldviews of many designers today, are profoundly interrelated. The best formalisms have always known this implicitly, and the recent ones ignore the reality of this relationship, and its challenges, at their peril. The present project reminds us that, before the formalisms of the recent post-historical idiom captured the popular imagination, Eisenman's was a formalism still to be reckoned with, both politically and philosophically.

A memorial-monument in the post-historical world must seek to function as an historical *machine*. Even the stillness and quiet that these machines strategically produce are not static: they are rather generators or dynamos of slowness, of *ralenti*. In direct opposition to statuary, environmental monuments appear historical precisely to the degree that they are abstract, entropic, geometrical. Of course it is we who invent the content that fills what has been cleared, for in the presence of these types of works we can do no else than to fill in and invent, to tell stories, stories whose very invention and telling actively mine the archives of collective experience. Extraordinarily, architecture today is everywhere beginning to take on much of the social role and discursive responsibility that in recent years has been abdicated by art. As art has given itself over to media, entertainment, and markets, architecture is beginning to appeal to intellectuals as a site of public debate, of new ideas, and of resistance to regressive social processes. Architecture, it may be said, is the site of encounter where geometry *meets* history.

So what type of story is generated by this memorial machine? One is struck at the outset by the lucid and natural reappearance in the project of the Eisenman use of the grid. The first story, the one that brought the grid here, to this politico-aesthetic juncture, is clear. The grid emblematizes the peculiar system of rationality that can be said to distinguish the European (Nazi) Holocaust from other acts of historical genocide, both those before and after it. The grid further represents both the legendary banality, as Hannah Arendt characterized it, of both evil and reason, and of their strange and incompletely acknowledged kinship, and the relentlessness and systematicity with which the campaign of extermination was carried out. The grid reminds us that the historical work of emancipation is not at all implicit in rationalist enlightenment as its good fathers claimed, rather that techniques of enlightenment and knowledge may easily, and at any time, be put at the service of atrocity. In the Nazi past, we are reminded, evil was fully routinized in reason and technique, worked out and transmitted through entirely standard and rational channels of instructions, directives, quotidian bureaucratic process, and industrial research and technics. In its directness and bluntness, the grid represents a system of rationality pushed too far, a rationality gone out of control, become monstrous. Whence the excessive and infinite aspect of the monument's site: what becomes palpable is the unstoppability of a system that, once put in place, assumes a life of its own.

The normative rhetoric of the grid permits us to experience this rationality as no other than *our* rationality, and in so doing endows the monument with a greater generality: it becomes an emblem of the dangers of twentieth-century rationality itself, a five-acre warning that must be firmly placed, and must remain, within our urban and psychogeographical midst.

The monument is said to resemble a cemetery, but more poignantly it may be said to resemble a City, the Twentieth-Century City, reduced now to cinders (the concrete is literally earth rendered barren and homogeneous by chemistry). The ruined City represents not only the specific threat to world Jewry, but the more general one to *civilization* itself, that is, to cosmopolitan life, to culture—all those things that to the genocidal mind Jews had come to represent. The Jew was a harbinger of the modern, he was the one whose very existence was rooted and transmitted in *ideas* rather than in the continuities of blood and earth, and as such he became the official scapegoat for the cosmopolitanism and the modernism that the Nazis simply could not bear. The single granite pillar amid the sea of concrete represents a kind of ember among cinders: if not the triumph of sanity, then the indomitability of conscience and ideas, and the emergence of both Jewry and cosmopolitan Europe from the fires of extermination after all.[2]

At the end of the first round of the 1997 Holocaust memorial design competition, the German authorities, impressed and intimidated by the rigor of the proposal, explicitly requested to be told how such a project could be explained to children. It would be unproblematic here to capitulate some realism to the children by explaining that what they have before them is indeed a cemetery, though one that has actually been opened up, permitting one to walk among the dead below as through a catacombs. This is a landscape, one would tell them, a city from which all life has been removed, where not a tree or blade of grass is permitted to grow, a desert, in the moral as well as the real sense, into which the Nazis had sought to lead European civilization. But the children would also need to be told that as a (typological) structure, the horizontally oriented memorial represents the *opposite* of a church with its manifold of vertical relations, a place from which, for a time during their own grandparents history, God seemed perilously absent. (This could be compared in counterpoint with Berlin's brooding and not at all ineffective *Gedächtnis Kirche*.)

These comments are admittedly heterodoxical yet they serve to stimulate middle-class complacency to confront issues about space, sacredness, and other, non-Christian and, more explicitly, immanentist traditions. This Holocaust memorial, one could tell them, represents a vision of the end of the world: it depicts a world of cinders, darkly representing the orderly and rational endgame urbanism of the camps. The Nazis, parents and teachers will be able to explain, believed they could eliminate from society those they did not like, and their efforts to do so were so savage and stupid that it resulted in nearly killing their own souls and their own world in the process. Thus the field of the monument also represents Germany as she left herself after the war, the result upon herself of her own provinciality, blindness and hatred.

The Ruin, as Walter Benjamin taught, is a corollary of the allegory, reflecting the submission of all things to the existential action of historical time. Eisenman has spoken of the memorial using the trope of crystals dissolving in water. This delivers to us not only an image of Ruin, but a powerful allegorical matrix as well. Just as there is no Benjamin without the determinant presence of Proust, there is no Proust without that of Bergson. It was Bergson himself who proposed the example of sugar cubes melting in hot tea as an example of the irreducibility and indivisibility of true duration. The seeming apotheosis of Time in these modernist Jewish thinkers reasserts a more fundamental and archaic aspect of the Jewish deity as the god of time and not of space (as in the Christian tradition). All of Jewish metaphysics can be said to be steeped in this principle, not least of which its central concept of the Sabbath.[3] Postwar aesthetics saw an overwhelming shift away from the metaphysics of verticality, embracing a centerless, beginningless, and endless horizontal field theory, which provided a far better metaphor for Jewish space and the disbursive distributions of the Diaspora (perpetual exodus. . .). The geometry of Eisenman's Holocaust memorial is a purely and simply bifurcating one: it announces its rationality only within this broader regime of ruin and allegory. Here it is the entropy concept that serves as the material equivalent of allegory: a physical repository of indivisible historical truth. While in its extreme form entropy represents the reduction of a system to homogeneity, a homogeneity that leads invariably to its death, it is more generally a measure of the difference in a system, a difference which is the dynamic of all life.

Spiral Jetty, Robert Smithson, 1970 (photo taken in 2002). Arguably the greatest artwork of the last century, Smithson's Jetty has just begun to realize its full temporal dimension as it marks the first beats of the cosmic rhythm that was designed for it by reappearing periodically within the ebb and flow of hydro-geological time.

The waves of (Z-axis) fluctuations that move through the rational grid are affirmations of this difference and life. It may be said that in the physical universe entropy connects all things, and certainly all things *to themselves*: every present expresses an algorithm in which are embedded the past and possible future states of a system. The Eisenman memorial rejoins the seminal themes of modernist Jewish thought through its anti-monumentality (recall the Jewish prohibition against graven images) and its affirmation of the continuity of time, even if this involves the perpetual corruption of forms. No regime for whom the finality and transcendence of forms is placed at the ideological center could help but see "Jewish subtlety" and temporal thinking as an enemy. Einstein's space-time relativity was thus necessarily condemned by the Nazis as "Jewish science." The present memorial also represents the death of German culture that was an inevitable result of the Jewish expulsions of the 1930s, and of the remarkable and unexpected return of Jewish cultural presence to Germany in the 1990s.

The field of integers that one beholds when overseeing the memorial from above (taking in its top, or broken, surface) powerfully resembles a torn text fragment from a sacred biblical scroll. Hebrew lettering possesses a remarkably three-dimensional tectonic quality, more so than either Greek or Latin script. On parchment scrolls these integers appear like perforated blocks, ruins, fallen walls. The Jewish cult of ruins comes immediately to mind here, especially that of the most celebrated Hebrew monument of all, Jerusalem's Wailing Wall. A field of ancient Hebrew text, gathered and aligned into ideogrammatic blocks, also produces the strange effect of flickering life—the play of solid and void, light and shadow, white and black, dot and line—playing itself out amid the spoils of time; that is a central feature of the Berlin memorial.

The Eisenman memorial, in its anti-, or displaced, monumentality, belongs more properly to a fluid and maritime tradition. In this long and established tradition, *the sea itself is given as a monument*, whence the nautical convention of the *burial at sea* that relies upon, and leaves, no fixed marker or rigid place. The sailor becomes one with—that is, indistinguishable from—the endless and boundless expanses of ocean. The memorial is therefore a single monument, *not* a collection of individual ones, and should not be mistaken for such. (It is composed of two entirely sufficient gestures: A. the setting into propagation of two undulating parallel surfaces, one solid and positive—the bed—and one "negative"—the "broken" or punctuated top sheet—and B. their topological transformation—the porous, reticulated space between.) It is a regrettable aspect of the design that the Eisenman office occasionally ignored this point themselves, incorporating into the design a random tipping of the individual plinths. This can only add an unnecessary narrative pathos to the project and reinforce the trite mental image of an old world cemetery ravaged by neglect. The project rather needs to emphasize its status as an elemental horizontal *landscape*—again a sea, as in Kant's example of the epic sublime, the sublime now but a remnant itself, and no longer achievable thanks to the midcentury death of the world, the sea too now in "ruins." Such an image poses a prickly challenge to the formal conventions of State ceremonies: the need for headstones, entrances, central elements, hierarchical structures, and master features. The final work, I believe, should eschew these principals completely, offering itself rather as does the "bow" or prow of a ship to a christening ceremony. This work should change the nature of the official state ceremony (it would in no way diminish it), either in keeping with the traditions of maritime custom or in the direction of some other alternative practice: wreaths might be laid out "to sea," or symbolically distributed through the site like cinders of cremation in the ocean. At present the work contains no center, no goal, and the rigor of this gesture should be scrupulously preserved.

Truer and more consistent than the idea of the tipping of the plinths is the commonly elicited metaphor that sees the Eisenman monument as a wheat field. The wind of history (contingency) becomes the corrupting/shaping element here, the source of a (one hopes) *non*-random deflection (bending) in the orientation of the plinths, one representing the over-rational and vain will of misguided human striving. The implicit argument here also bifurcates neatly, and should be clearly accentuated. Formally, what we have is a *fluid* and indeterminate movement coursing semi-freely through a grid; in fact, not just a movement but fully an existential ethos that embraces but does not rigidly resist the multiple action of time upon matter. This not only allegorizes the democratic movement of the dissolution and merging of transcendent (fascist) State power ever more inseparably into the body politic, but also represents the cultivating, deepening and softening effect that memory works on any human civilization that permits itself actively to incorporate its effects.

The Eisenman Holocaust memorial represents an ethos and a set of arguments about civilization presented to a culture and to a vast set of viewers by means of a pure and elemental geometry that for all its rigor is never harsh. It departs from convention through rhetorical and philosophical necessity and never simply to achieve histrionic formal effects. In December 1998, German philosopher Jürgen Habermas sent a letter to Peter Eisenman in which he implored the architect not to make any programmatic concessions to the German authorities. According to Habermas, the goal of the German officials would be to achieve a compromise on the existing scheme, to replace the brute importunateness of a public architectural fact cast in concrete with social "software"—an educational center and program "about" and "on" the Holocaust, etc. Habermas's point—one of the great recent testimonials to the robustness and resistance of architectural facts—is that if the memorial were built in its proposed form, history, for a long time to come, would be unable to elide, revise, or reduce the force and content of the structure's embedded messages. The work would produce a forced and sustained *anamnesis*, as it were, at the public, urban, and historical scales. The official alternative, a mere bit of fluid institutional programming, is always subject to change or suppression on a moment's whim and according to the vicissitudes of history. What is critical, Habermas pointed out, is that memory should be projected into the future in concrete, irrefutable structures, as something that cannot be absorbed, but must be confronted again and again within changing contexts—stubborn as hell yet ever anew.

To Habermas's important argument must be added a small addendum. The Eisenman Holocaust memorial proposal has already undergone considerable revision at the request of city planning officials, citizen groups, and ministerial pressure. With but one exception these changes have served only to trivialize and normalize an original gesture of great power, clarity, and integrity.[4] True, the proposed memorial is a stark and stern structure, but it is exquisitely economic in its expressive means, powerful and serene and controlled like few works realized anywhere in this century. The historical realities and the conditions it recalls are not happy ones, but neither are they exaggerated here to create a spectacle or amusement of any kind. It is a calm work that will produce nothing but wonder and introspection and spirited debate. There is nothing sad or morbid or bleak about it, and there is no reason to domesticate it with platitudinous planning gestures such as shrubs and saplings, friendly outline shapes and smooth playground and sidewalk transitions. I personally know of no monument as powerful and lyrical as this one, as direct and as elegant in its language and rhythm, and as appropriate to its serious subject. It should remain a work of historical and philosophical introspection first and a people's park second. These two functions are by no means antagonistic, but the triviality of the second function must not take precedence over the first. The skateboarders, the orators, and the hide-and-seek players will come, and should be encouraged. Yet the integrity of an historical gesture of great proportions is at stake here, and the short-term populist demands and crowd-pleasing fears of politicians must sometimes take a back seat to inspiration and to a city's and a culture's aspirations to greatness.

1 After considerable political discussion and many design changes, the Memorial to the Murdered Jews of Europe was completed in 2005 and dedicated May 10, 2005, eight years after the original proposal, by Eisenman with the artist Richard Serra, won a 1997 invited international design competition.

2 The anomalous pillar of different material and aspect randomly situated in the field was not in the end included in the built memorial but figured at one moment in the project brief.

3 See Abraham Joshua Heschel's essay, *The Sabbath: Its Meaning for Modern Man* (New York: Farrar, Strauss and Giroux, 1951).

4 The one valid addition to the original scheme was a request to build a Holocaust archive building directly opposite to the monument. The Eisenman proposal for this building was a work of impressive restraint and elegance, a gesture so clean that it seemed raw, so self-complete that it is almost totally visually silent. *Gravitas* has not been a characteristic of the work of the Eisenman office, and this work is a notable exception.

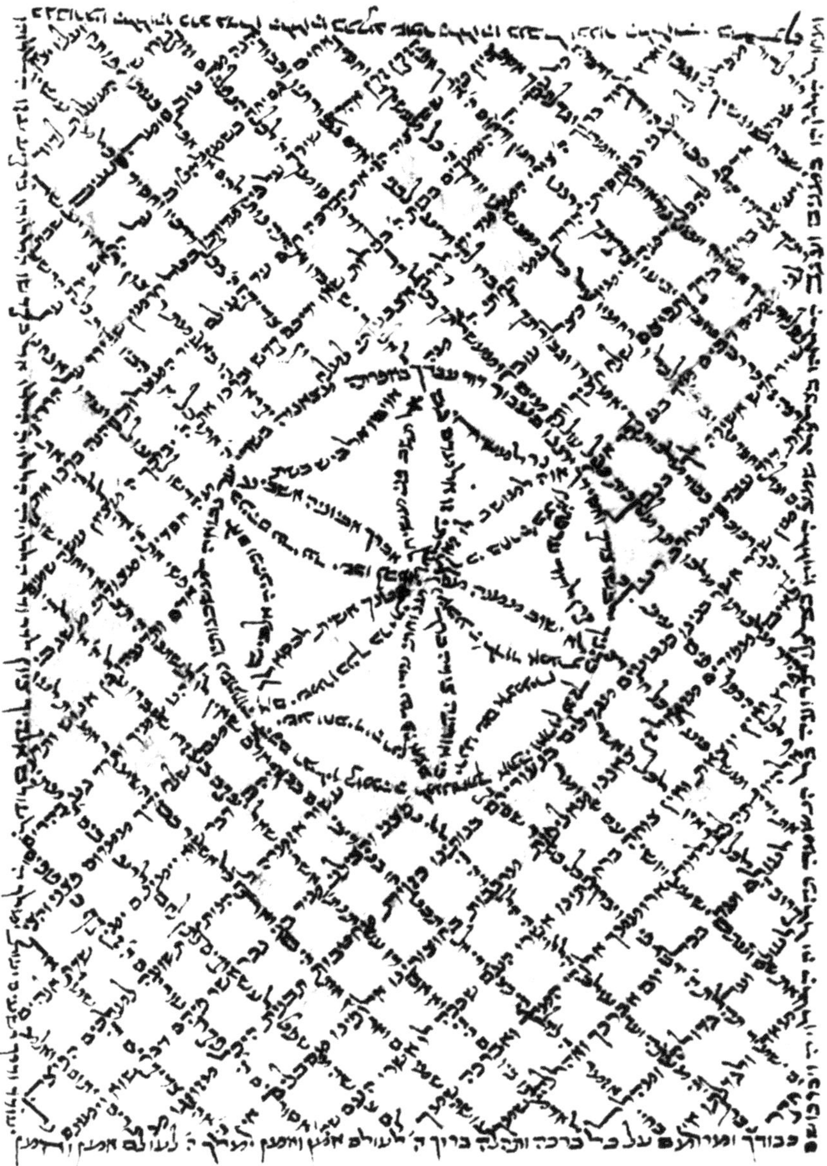

Micrography. Within the Hebraic interdiction against graven images, text bears the weight of mystical transformations of line.

FORMAL, INFORMAL AND THE JEWISH QUESTION

No principle marks Hebraic civilization more deeply than the interdiction against graven idols. The Jewish god Yahweh, himself faceless, chaotic, and mercurial, formed the world out of, and appeared to his subjects only through, language (this to the profound anxiety and perplexity of Moses and Abraham). The Elohist god eventually replaced the idolatry of static figures (the recidivism of Aaron) with the fluidity, the *performativity*, and the infinitely flexible rigor of a textual practice (Tablets, Torah, and Commentaries). The historical Exodus of Jews from Egypt was itself an act of multiple decomposition—the release of free vectors onto a Steppe (for only forty years in the Sinai were sufficient to permit the formation of an entirely new human "type")—and the severing of relations from a despotic State formation (subjection to pharaonic sovereignty and its tectonic world of closed monumental forms). Judaic space—to extend an hypothesis of Bruno Zevi—can be understood only through its nature in the *episodic* and therefore the plastic, and this in radical opposition to the solid forms of imperial power (the pyramids). Though the birth of the Hebraic tradition constituted a breach of decisive historical novelty, it is no secret that it was the Egyptian, not the Hebraic, ideal that was later reembodied in what came to be known as classicism.

The regimes of later Athens and Rome, and their subsequent—corrupted and fetishized—reception from the Renaissance to the present, provided the framework for a blind cult of order (understood in the most impoverished and oppressive sense) based on bureaucratic values of symmetry and proportion. Classicism is the formalism of empire, the sleazy-grandiloquent expression of a hidden will to purity, an *idolatrous* bulwark against the genuine forces of historical process: the mobile, the unrooted, the destabilizing, the new. To be sure, other formalisms exist, *informal* ones for example, though these are typically said to come from an outside (of course, only despotic formalisms create the illusion of serene, static, ahistorical insideness).

In our own century these forces of innovation have intensified (modernization) even if they have maintained a millennial mythical characterization: adventitious Hebraism. The anticlassical formalisms of Einstein (the mollusk of space-time) and of Schönberg (dodecaphonic and timbre-composition) revolutionized the absolutist "orders" of Newton and Bach (on rigorous mathematical grounds!), yet these were duly abhorred and denounced as "Jewish science"

First published in *Progressive Architecture*, February 1995. This piece was written in response to the question posed by Peter Eisenman: "Can form be defined politically?"

MARINE SHUTTLE
Murakami
バス

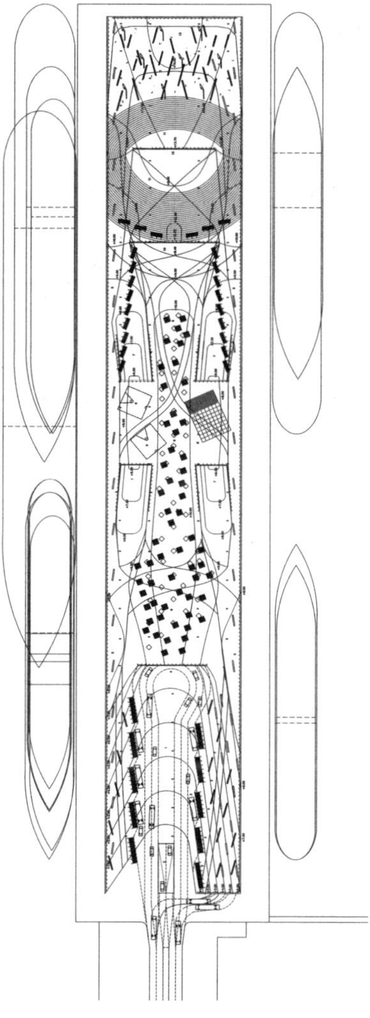

ANIMAL MACHINE

Arnold Schönberg's *Moses und Aron*, Berlin, 1959.

and "cultural Bolshevism." Perhaps because Hebraism belongs to the openness of *diaspora* (*dispars*), to free action, and to actualization through performance in an essentially *interpretative* landscape, it has often been assailed by the fascist-classicizing mind as the result primarily of pernicious "influence," rather than "work."

The fear of fluency, and the inability to tap the transformative power of the turbulence which issues from it, are the sinews on which classicism is built. All Hebraism enfolds within it the volatile powers of the informal formalism of manifold time. One need only recall Bergson's *élan vital* and Proust's anti-novel of spontaneous *Remembrance*. And who today can seriously doubt that Heidegger's disfiguring concept of *Dasein* is anything more than a fearful classicization of Being in a spatial *there*, that is, in a rootedness-in-place (*Blut und Boden?*) from which the wandering ("cosmopolitan") Jew is not only excluded by nature but is also made to represent the diametrical opposite: the very embodiment of the rootless, disruptive forces of change: technology (money and ideas) and modernization. Yes, in the West, the genius of transformation, innovation, and revolution is by no means always Jewish, *but it is always Hebraic*. For what else is Western history if not the continuous, varied unfolding of the archetypal strife between the biblical brothers Moses and Aaron (as Schönberg, prophet of anti-classicizing dissonance, subtlely showed in his opera), between the Hebraic god *of events* and the Egypto-Hellenistic one of idols, orders, and repressive forms?

To the question, "Can form be defined politically?" we can only answer, "When has it not been?"

FOA's Yokohama International Port Terminal contains no right angles and no surfaces that separate or detach from the continuous four-dimensional flow of matter and void that constitutes the building. While continuums of this type are typically associated with hydraulic, or more generally, *rheological* systems such as those found in foams, lavas or living tissues, the present ones are clear products of the digital modeling revolution that swept design milieus in the late 1990s. But this last phrase, so common in contemporary architectural journalism, carries little explanatory power, with respect to either the form of the present work or even to the idiom of which it partakes. The early work of FOA is profoundly rooted at once in the social Modernism that characterized European design of the 1990s and in the American organicist movement that even more broadly affected design practice in the public eye of the last decade. The Yokohama Terminal has the arguable distinction of being one of the only—and certainly the first—major works of neo-organicism to exhibit even a small part of the rigor and promise of the theory's underlying ideas.

The intellectual revolution at the root of this organicism can be found in the postwar era, particularly in the work of Norbert Weiner and the extension of field equations to general and intractable phenomena of highly correlated systems containing many parts that rely on signal-processing to coordinate their global action and form. Out of this worldview came the first "sciences" that connected living and machinic systems into potentially explanatory wholes.

Among these, the Yokohama Terminal makes central use of (without declaring) sophisticated geometrical-mathematical models (best known in the 1970s as "catastrophe theory") developed by the mathematician René Thom. Thom's interest was to see whether many of the form-emergence events that characterize living or biological structures could be formalized or explained with mathematical models. Among his startling discoveries was an explanation of the bifurcation in the embryo that results in the differentiation of initially neutral tissue into either a male or female genitalia. This "difference" is the outcome of a single variable, and whether its value presents itself as greater or less than 0 when mapped on a plane. The astoundingly simple diagrams that Thom generated are the foundation of his theory of tissue bifurcation (types of opening in unfolding surfaces) and, in the Yokohama Terminal, become the source of a whole extended "parti" of splitting and converging surfaces.

By refusing to impose a form at the outset, but rather to let it unfold as if in some type of perpetual embryogenesis out of an initial undifferentiated state, FOA achieved an astounding and unprecedented effect: a building that appears more as a thing born than a thing made, one in which its informal, "pre-informational" state is as legible and effective as its formal and articulated one. The result is to offer, in a single apprehension, a marriage of dry and wet computation in a single object—a perfect analogy for an ambitious work whose primary concern is to resolve and connect (without suppressing) the nautical-land interface.

ROOM & BOARD
ROOMS

City as Predator: *Crossfire*, Edward Dmytryk, 1947.

Houston, Texas.

NEW BABYLONS: URBANISM FOR THE NEW MILLENNIUM

First published in *Assemblage* 25 (1995).

The sinews of modernity may be discerned in three succinct and interwoven social processes of nearly identical origin: the processes of rationalization, of industrialization, and of the more concrete process into which the latter two are subsumed, that of *urbanization*. Yet these forces are all too readily associated in the historical mind only with the specific upheavals of the late eighteenth and nineteenth centuries, and too rarely with the more bizarre and idiosyncratic developments of our own contemporary world. Indeed a great deal of our intellectual life during the last thirty years has actually sought to minimize awareness of the intricate and increasingly subtle continuities that these fundamental social forces represent through their stubborn, if increasingly invisible, persistence. The recent bankruptcies of two important but obfuscating intellectual movements—the so-called "postmodernist" philosophy on the one hand, and on the other, the retreat within our plastic traditions to increasingly hollow formalisms of "style"—are a sign that a window is now opening up through which we may again apprehend, perhaps with a new clarity, the systematic play of forces that has always driven the historical transformations of social, economic, and mental life. The study of the results and implications of these processes, of their movement, strife, and intermixing, is what is increasingly being understood today by "urbanism."

The term *urbanism* is admittedly used here rhetorically: it deliberately summons back into existence a type of textual practice that subsists today only in the frailest form. Indeed, what we need today is to revive the urbanism of the essay (discursive or graphic), the urbanism of the speculative historical or philosophical treatise, in a phrase, the semi-"delirious" *urbanism of ideas,* as a vital alternative to the present ultra-narrow, over-professionalized, sterile urbanisms of the clerical disciplines. No longer content with the milquetoast urbanism of merely remedial design propositions, nor the camp urbanism that transforms the advancing, often savage deprivals of the modernization process into cult objects (i.e., suburbs, nets, edges, and spectacle), what we need is a *genealogical* urbanism that both invents and unearths embedded histories-in-the-making, and through such invention transfigures and transvalues the very landscape on which it operates.

An exemplary text of this type is "Stim and Dross" by Lars Lerup.[1] Lerup's text might best be described as a "paraliterary" work (in the sense that this term has been applied to the later work of Roland Barthes), a text that functions as a type of early *portolano* of the late century metropolis in that it sketches out, with no pretense to exhaustive projection, a provisional series of actantial characters (generative dramatic structures); fixes privileged points of urban threshold and rupture where affects naturally cluster in the landscape; establishes an inchoate lexicon of the increasingly abstract and *creolized* objects (i.e., initial "corruptions" that emerge into primary use) that punctuate the new urban field; and most important of all, supplies the above elements with a set of algorithms that link them into a mobile syntax of procedures, routines and events. The text is not only notable for its expository style (it seems an almost promiscuous amalgam of J.G. Ballard, William Gibson, Reyner Banham, Jean Baudrillard, even Carl Sagan and Lerup's career-long guide, Michel Foucault) but also for the unity and originality of its analytical machine: Lerup's Houston is a marvelously integrated metabolizing *beast*, an animal rife with tropisms, habits, appetites, exudations, cadences, and transient lifecycles. The mysterious and complex "stimdross" is, moreover, an *active* concept that targets the city-object obliquely as a fulminating *ecology* of multiple forces now inhabiting *orders of time*, and not only orders of space. It gives place to what is certainly the essay's most important concept: the principle of the *megashape*, a dynamo of spontaneous continual efflorescence not unlike the orgasmic cloud of Marcel Duchamp's untouchable Bride—self-generating, evental, entirely organismic, beyond all totalizing grasp, yet pure solid geometry, and nothing but.

Lerup's text, a kind of "New Babylon" for the Metropolitan Age, might seem an eccentric work, though only in relation to the most diffident, bureaucratic forms of urbanism that, in the last two decades, have become the norm. In reality, its spirit belongs to another confident if sporadic tradition that includes the work of Lewis Mumford, Reyner Banham, Paul Virilio and Rem Koolhaas as much as that of Alexis de Toqueville, Georg Simmel, Max Weber and Walter Benjamin.

The new "urbanism" I am calling for recognizes that it is still *art forms*—those of the cinema and of literature—that have provided the most compelling urbanist archive produced in the 20th century, and it is to their oblique and infinitely rich documentary spirit, and to their examples as labors of the imagination, that we will need to adhere.[2]

1 See Lars Lerup, "Stim & Dross: Rethinking the Metropolis," *Assemblage* 25 (1995), pp. 83–101.

2 A wave of urbanist projects emerged after the writing of this essay in 1995, including the Metapolis group in Spain, the Multiplicity group in Italy and throughout Europe, the Harvard Project on the City, the photodocumentary school around the Berlage Institute in Rotterdam, etc. Sadly, most have now faded away.

Some years ago, in the wake of a brutal attack on a jogger in New York's Central Park, the *New York Times* published what could only be called a deliriously obscene—though also strangely beautiful—map of the crime's site showing two converging vectors of entirely disparate character and movement. Though the result of the meeting of these two lines produced an outcome of great ugliness, the subsequent police interrogations and *New York Times* reportage brought for the first time a new formal and social substrate to the public urban imagination: the concept of "wilding."

A young gang from an East Harlem ghetto had allegedly formed and descended like a sudden, spontaneous tsunami wave on New York's multimodal, multiple use, and prodigiously democratic urban "steppe"—Central Park—to wreak havoc with whatever crossed its path. But it was a startling and unexpected path, one that followed none of the prescribed and designed routes through a landscape, even Olmstead's organic, picturesque one. The *Times*' diagram displayed two very distinct modes of holding and carving space: that of the white, middle-class jogger adhering to prepared pathways of program and roadway; and that of the wilding (their own invented term) posse, ranging freely—and freeformly—through forest, darkness, and plain, encountering all manner of phenomena: monuments (to deface), natural features (bogs, ponds, thickets, and hills to negotiate), settlements of park squatters (to engage, intimidate, or disturb), etc.

The violent encounter of the two worlds, stockbroker/office worker vs. gang member/street dweller, that of the white, bourgeois Upper East Side and that of the besieged and socially disenfranchised ghetto, was nowhere more dissonant, and dramatically displayed, than in their programmatic deployment of space. The *Times* map was aberrant, and it shocked. Among its most obvious and singular features: how rarely the wilding trajectory encountered roadway; not only did it neither follow nor seek existing circulation paths, it seemed oblivious to them, heedless of all those "Other" patterns and processes that belong to the organized, optimized (white, establishment) urban world.

Wilding became a new and terrifying word for urban drift; for ad hocism; for the collective, unstable phenomenon of pack, mass, and crowd; for the spontaneous emergence of epidemia and "stim;" perhaps most significantly, for the unruly and uncontrolled emancipation of self-organizing social forces from the rigid geometries of socially and behaviorally engineered urban space.

WILDNESS
(Prolegomena to a New Urbanism)

***

America, one might have thought, would already have learned the lesson of wildness.[1] The Vietnam War was a case study in the futility of the classical Western strategy of holding space. The almost total super-fluidity of the Vietcong guerillas moving in and out of tunnels, changing sides and allegiances with every fall of the night, transforming the concept of "frontality" and "battle" by mastering a logic of the ambush, booby trap, and disappearance, made mincemeat of American techniques of waging war on a schedule and within a numbered, geographical grid. (It is a reflection of this same logic—the intuitive capacity to move fluidly between tactical and strategic modes in relation to fluctuating conditions—that computers may well never manage consistently to beat the human master at chess.) The Vietcong saturated the landscape and jungles, and even when they weren't literally present at all, the fraughtness of their threat persisted; indeed, they succeeded to "jungle-ize" even the city of Saigon. They did this, and prevailed, because they fought a *tactical* war in time, and so *strategically* rendered themselves unfindable in space.

***

The "wild" is the logic of animal societies (packs, flocks, and swarms), of the immixings and inadvertencies of the natural world (storms, quakes, abundance, extinctions), and of complex adaptive systems in general, even those of an entirely artificial kind. Since the mid-1980s Santa Fe, New Mexico, has become a main center for the study of such systems. The phrase "fast, cheap, and out of control" has become an unofficial slogan of scientists and systems designers working at the Santa Fe Institute for the Study of Complexity. The concept behind the slogan suggests that extremely intricate systems can most effectively be built up *messily*, in steps and layers, from approximate rather than finished and perfect parts, and incrementally over time, rather than in one fell swoop of assembly. *Indirectness*, it appears, is actually the secret to achieving a robust, adaptive, flexible, and evolving design. Such systems, until now, have been essentially self-designing. They are wild systems that range and explore and mine their environment, that capitalize on accidental successes, store them, and build upon them. They are densely layered systems, in which every layer functions and contributes—though in a

Red-billed queleas, Okavango Delta, Botswana. The most numerous birds on earth, queleas travel in huge clouds when they leave their roost, forming great synchronized moving architectures while on the go. When amassing on the African horizon, they are frequently mistaken for grassfire.

totally unknown and untrackable way—to the shape, behavior, and stability of the whole. There is no central control, and the "design" does not come from the whole and trickle down to the parts, but rather travels in the opposite direction. Such bottom-up systems are called "subsumption architectures." The natural universe is itself a subsumption architecture utterly "out of (central, single) control." To give place to such out-of-control, adaptive, robust, self-directed designs is to allow, or to install, a degree of wildness within them.

***

No real historical space—be it Central Park, the jungles of Vietnam, or, indeed, the new abstract modeling spaces being developed at Santa Fe—is ever univocal but always multiple and intertwined. This multiplicity and embeddedness, combined with the critical presence of a "thick" time, allows for the open-ended interactions of parts—the hybridizations, blendings, and conflicts—that effectively destroy a structure's determinism and that feed its wildness. (Wildness emerges in a system once we lose the ability to predict—from the outside—what it will do.)

This category of structure (which includes at least some architectures and most cities) can be thought of as *polygenetic* structures. One characteristic of such structures is that they are large—though in the qualitative and relative sense more than in the absolute. They are large in that they are complex (generated from an indeterminate number of distinct sources, distributed through many, or *n* dimensions), and fundamentally open (i.e., wild) and unfinished. The *urban* may be characterized precisely by this constellation of "polygenetic" qualities; it is a promiscuous, evolving manifold, to be distinguished whenever possible from the term "city," which, in its saddest form, may consist of nothing more than a stillborn, mechanical, man-made design.[2] The urban, in other words, may and often does exist at the scale of a piece of architecture, and conversely may fail to exist in city-sized agglomerates. Its existence depends only on whether a "computational" threshold has or has not been crossed. Approaching this computational threshold as a source or reservoir of work, information, and form is one of the central challenges of contemporary science, economics, and design. Over the next two decades the amount of urban substance that has existed on earth throughout all its

Timber wolf pack.

history will double. Less than five percent of this will be designed or planned in the traditional manner. The rest will follow one-dimensional pathways of the market. Today's task is to induce—dare one say *grow*?—these polygenetic structures, or more accurately, to program the systems that enable such structures, in their turn, and at a sufficient wild distance, to assemble themselves.

Evolutionary process—the formation of organic meshes—was always built into urban matrices as the result of historical pressures and patterns embedding themselves into matter, in an open process over time. Design today must find ways to approximate these *ecological* forces and structures, to tap, approximate, borrow, and transform morphogenetic processes from all aspects of wild nature, to invent *artificial* means of creating living artificial environments. We must learn to see design algorithms everywhere we look. In time, we will earn the right to call ourselves urbanists again.

1 Perhaps one of the clearest examples of the effective assimilation of lessons of Vietnam can be found in the John Badham's film *Blue Thunder* (1983). In Badham's film, Vietnam technology and experience are deliberately appropriated by the government in the service of a comprehensive recoding of urban space away from linear paths and punctuating objects in the landscape in favor of a totally fluidizing and wilding configuration made possible by helicopter mobility and optical and auditory surveillance. An entirely new type of violence and mastery are made possible.

2 If our culture has lost interest in "the city" it is because we no longer know what it is (or whether it exists at all). But this is no license for complacency or passive acceptance of what has come to replace it: exurb, garrison communities, middle landscape, edge, etc. The urban is the primordial modern human wilderness, currently under siege by laissez-faire ideology, and the new Republican-led feudalism that is striving to shape it. It is one thing to let go of the classical city, another entirely to abandon the emancipatory notion of the *urban*.

Central Park, New York.

This book was set in ten separate typefaces deployed in sequence through the volume. The transitions occur at set intervals that correspond to signature divisions in the binding, specifically at points where gatefold pages are inserted. The fonts were organized into two different families, those with stroke weights that vary— Centennial, Melior, Excelsior, Egyptienne, Charter—and those whose stroke weights are uniform— Glypha, Serifa, Rockwell, Caecilia, and Egyptian. The purpose of the first sorting is to establish a register within which to vary and modulate the "gray scale" of the page (the proportion of black figures to white ground). The individual fonts were then organized according to their internal proportions or the relative sizes of their uppercase and lowercase forms. Type sizes were slightly modified to "standardize" these proportions and to smooth out the typographic transitions. A third sorting process was then applied to determine the order of type widths so that the book begins with the tighter fonts and develops or proceeds to the wider ones. By controlling the gradients and interrelationships of these three properties a subtle effect of "change over time" or propagation is achieved in the flow and density of the book, much as heat would travel, or dissipate, when moving through a solid.

The purpose of distributing these disjunctive events across the book is not to conspicuously violate the fundamental rules of book design (a fashion of the 1990s coincident with many of the themes treated in the essays) but to register variations and disturbances in the book's flow and equilibrium, in particular, in the relations between the major keys (the main essays) and minor keys (the foldout essays). The typographic registration of turbulences has been engineered to produce a distinct perceptual effect on the reader yet to hover below, or just at, the level of conscious awareness.

A NOTE ON THE TYPE

IMAGE CREDITS

Front end paper: AP Images/Luke Frazza/Pool
20–21: *Cell Phones*, Atlanta, 2005, by Chris Jordan
22: Library of Congress, Prints & Photographs Division, The Work of Charles and Ray Eames
24: Courtesy of Hibernia Management and Development Company Ltd.
28: Picture used with the permission of The MITRE Corporation. © The MITRE Corporation. All rights reserved.
32: Photograph by Paolo Valenziano (www.azhim.com)
36: U.S. Navy photograph by Ensign John Gay
42: Richard Meredith Hardy/Rex USA
48: *Game of Life*, John Conway (1970)
50: Courtesy of Beat Engel
52: Courtesy of North Sails Group
54–55, 58: Courtesy of the Estate of R. Buckminster Fuller
62–63: Brent Finley (www.brentfinley.com)
66–67: Courtesy of NASDAQ
68: Courtesy of NL Architects
70: Courtesy of MVRDV
74: *Manhattan Office Building*, Andreas Feininger (1962)
82: Courtesy of Greg Lynn FORM. Photograph by Jan Staller
90–91: Courtesy of IBM Archives
94, 96: Courtesy of Patek Philippe
99: Courtesy of The Smithsonian Institute
100: Photograph by Kazys Varnelis / AUDC (2002)
106: Library of Congress, Prints & Photographs Division
108: Reuters
118: Bureau of Inverse Technology
126–127: Courtesy of Human Genome Project
130: Courtesy of Richard Smalley/Rice University
134: Film still from *Playtime*, dir. Jacques Tati (1967)
136: Courtesy of Vito Acconci
140: Courtesy of Sonnabend Gallery
144–145: Dmitri Kessel/Getty Images
154–155: Courtesy of Josef Astor. Originally published in *Vanity Fair*.
158–159: Photograph by Blaine Harrison
164–165: Courtesy Roland Halbe
170–171: IKONOS satellite image courtesy of GeoEye, © 2006.
176: Courtesy of The Library of The Jewish Theological Seminary
178: Courtesy of the Arnold Schönberg Center
180–181: RKO Radio Pictures/Photofest
182–183: Photograph by George O. Jackson
188: Image © Peter Johnson/CORBIS
192–193: Vincent Laforet/The New York Times/Redux
190: Courtesy of Wisconsin Department of Public Resources
Back end paper: Picture used with the permission of The MITRE Corporation. © The MITRE Corporation. All rights reserved.

FOLDOUTS
18: Beat Widmer
38: Photographs by Hans Werlemann/Hectic Pictures
58: (Left) Courtesy of Cecil Balmond/Arup. (Center) Courtesy of Shigeru Ban Architects
78: Courtesy of Abalos & Herreros. Photographs by Bleda y Rosa (Maria Bleda and Jose Maria Rosa)
98: Peter Mauss/Esto
118: Photographs by Ramon Prat
138: Courtesy of Coop Himmelblau. (Left) Photograph © Gerald Zugmann. (Center) Image © ISOCHROM.com, Vienna
158: Photographs by Hélène Binet
178: (Left) Courtesy of Foreign Office Architects. (Center) Photograph by Satoru Mishima